Geometry by Discovery

DAVID GAY
University of Arizona

John Wiley & Sons, Inc.
New York ✴ Chichester ✴ Weinheim
Brisbane ✴ Toronto ✴ Singapore

COVER PHOTOGRAPH   Josef Beck/FPG International

MATHEMATICS EDITOR   Barbara Holland
MARKETING MANAGER   Jay Kirsch
ILLUSTRATION COORDINATOR   Jaime Perea
ILLUSTRATION STUDIO   EM Technical Art Services
PRODUCTION EDITOR   Ken Santor
DESIGNER   Madelyn Lesure

This book was set in Times Roman by Bi-Comp, Inc., and printed and bound by
R. R. Donnelley–Crawfordsville. The cover was printed by Phoenix Color Corporation.

Recognizing the importance of preserving what has been written, it is a policy of John
Wiley & Sons, Inc. to have books of enduring value published in the United States
printed on acid-free paper, and we exert our best efforts to that end.

The paper in this book was manufactured by a mill whose forest management programs
include sustained yield harvesting of its timberlands. Sustained yield harvesting principles
ensure that the numbers of trees cut each year does not exceed the amount of new
growth.

Library of Congress Cataloging in Publication Data:

Gay, David.
    Geometry by discovery / David Gay.
        p.   cm.
    Includes bibliographical references and index.
    ISBN 0-471-04177-7 (cloth : alk. paper)
    1. Geometry. I. Title.
QA453.G35   1998
516.2'076--dc21   97-15258
                                CIP

Printed in the United States of America

10 9 8 7 6 5 4 3 2

To Terry Mirkil, my teacher,
and David Kelly, my colleague and friend:
for helping me discover what's possible in learning and teaching.

Preface

❋ *To The Reader*

This book is a do-it book for geometry. People dance, paint, play musical instruments, compete in sports, repair automobiles . . . and do mathematics! In this book I have tried to provide you with many opportunities for doing geometry. I want you to have a real adventure. I want you to be swept off your feet. Like any human activity, there will be ups and downs, frustrations and joys. I want you to struggle with the ideas and the problems. But I also want you to experience the exuberance of being successful, of solving problems, and of gaining extraordinary insight.

To get you solving problems and doing mathematics right from the start, the problems and ideas have to grab you. The ideas have to be new (or old with new seasoning). But they also have to be ideas that you can grab on to, using the geometry that you know. Thus I have chosen the Leading Problems of each chapter to be understood and appreciated right off, without a heavy investment of time.

To keep you going, I change the vistas, more or less forcing you (!) to look at geometry with new eyes. This way you and I won't get bored or bogged down, shrivel up, and blow away. But geometry is more than just a series of events. There's additional spice when there are links between them and surprises occur when you come upon the links. I have planned for these, too.

As topics change so do their reasons for being. Some are there because you've seen them before; they can provide connections with your past while being relished in new ways. Others are there because they're useful; folk out there may want to know about them to make money, to save the environment, etc. Still others are there because they're natural extensions of what you know; yet others, because they're beautiful and part of the culture of the mathematical community.

The book is also set up to get you to carry out geometric activities using different approaches. Sometimes it will ask you to make a model or two. Other times it will suggest that you use technology (more on this below). Having several ways to approach a problem makes you a powerful problem solver.

At other times the book will suggest that you work on problems in a group of your peers. Geometry is a social activity. The more you talk about it the more alive it becomes and the more you learn. You can gain from knowing and absorbing a friend's different (and possibly, quirky!) insight. As you attempt to explain an idea to a friend, the idea will take on a new dimension in you. Your friend will always view the idea from a different field of experience than your own. This interaction will add richness to your experience of geometry.

Most importantly, I have chosen the topics and carved paths through them so that you can discover the theorems for yourself. I have chosen problems (and topics) that (I hope) will grab you. But after that I have left the solutions to the problems and the discovery of theorems up to you. I have laid out the book to minimize my telling you the answers and maximize the possibility of your finding the answers yourself. The book does not set itself up to be an expert for you to mimic. The book is a guide. It sends you off with a problem or a topic and suggests ways you can discover solutions and concepts on your own. Why do I do this? First of all, it's holistic: you will need to use all your facilities, all your senses to be successful. (It may be frustrating at first!) Secondly, the ideas you discover will really be your own. Thirdly, you'll be hooked. This is what mathematical research is all about. A student who used an earlier version of this book had this to say: "At the beginning, I was very frustrated with the lack of structure. You have made it easy for me to experiment, to decide where to start, where to go. This class is probably more real mathematics than any class I have had to date because dialogue, risk, experiment, doodling, playing are all the basis of discovery. Rigor comes after the insight."

After you have gone off, had your adventure (solved the problem), and come back to society, you are then required to share your story with your peers. You will need to become a good story teller. This also requires a certain amount of reflection on what happened during the adventure. You will need to learn how to reflect.

Finally, the book offers the possibility for further involvement in geometry beyond what is "covered" in the main part of the book. This comes in the form of a final chapter on projects. These are opportunities for independent study and research, either by yourself or with a group. This may give you a culminating experience with the course in which you can show off your involvement in geometry by sharing your findings with your instructor and your friends. In this chapter, you will find suggestions on how to communicate results through oral reports, papers and displays.

In the end, I want this book to

- enhance your geometrical intuition
- provide you with experiences to help you make sense of more formal geometric discussions
- empower you with new approaches to solving problems
- enable you to realize that mathematical techniques emerge from the solution to problems
- help you see that mathematical ideas are connected—the solution to one problem can be used to help solve other problems. (Sometimes this occurs through the use of analogy and generalization.)
- elicit gasps of surprise as you encounter these connections
- get you to discover mathematical ideas and solutions to problems yourself
- help you to have a great time as you acquire a taste for mathematical research (not necessarily at the "frontiers", but at a level appropriate for you)

✱ *Format of The Chapters*

An introduction sets the tone for each chapter and outlines the types of problems to be solved. The rest of the material occurs in the format of Leading Problem, Solution, Discussion of Mathematical Ideas, Exercises—cyclicly and roughly in that order throughout the chapter. Some of the exercises are routine, many are less so, and some are open-ended. Each of Chapters 2 through 8 ends with a section titled Full Circle, which highlights the important events of the chapter. This is followed by Notes—mostly historical comments on the material of the chapter. A section titled References ends each chapter. The formats of Chapters 1 and 9 are a little different. More details on all the chapters are given in the overview below.

After each Leading Problem you will find the icon **!** and something like "Stop! Try this before reading on!" This is a serious admonition. The idea is that YOU get engaged in the problem before reading what SOMEONE ELSE has to say about it. You may not solve the problem completely but you may arrive at a partial solution and you will gain insight. It's also likely that you will come up with an approach that is uniquely yours. "**Problems for right now!**" also call for your immediate action. These are problems that should be solved before reading on as opposed to the numbered problems which may be solved after reading the text.

✱ *An Overview of The Chapters*

The chapters differ in geometrical flavor, types of leading problems, and processes used to explore and solve problems. On the other hand, there are a lot of connections across the chapters among mathematical ideas, problems, and their solutions. Below are thumbnail sketches of each chapter, indicating some of the variety and relatedness.

- **Chapter 1:** This is an introduction to the kind of thinking and discovery to use and experience in the rest of the book. It contains a wide variety of accessible, yet challenging, problems and includes a collection of Mini-Projects for extensive investigation.
- **Chapter 2:** This is a fresh look at familiar ideas—area and volume—with special emphasis on making sense of the formulas for the circumference and area of a circle, the volume of a pyramid and the surface area and volume of a sphere. There is a lot of use of analogy in the investigations and arguments. There is some opportunity for making models.
- **Chapter 3:** Out of the familiar notion of two-dimensional polygon this chapter carves an analogous three-dimensional notion. There are many opportunities for making models and for using The Geometry Files.
- **Chapter 4:** This chapter contains some real world uses of geometry. It's the first of several with an efficiency theme (the other two are chapters 7 and 8). The leading problems are based on the question What is the shortest path? There are surprising connections with mirrors and conic sections and many possible explorations with Geometer's Sketchpad®.

- **Chapter 5:** The ideas from Chapter 4 are used to investigate kaleidoscopes. A lot of exploration results in the discovery of what makes a "good" kaleidoscope. This chapter exhibits many examples of powerful results that follow easily from simple assumptions and includes a surprising side excursion into spherical geometry. There are instructions on how to make a real kaleidoscope. The chapter contains many connections with tessellations (Chapter 1) and polyhedra (Chapter 3).

- **Chapter 6:** This chapter attempts to answer the question, "What is symmetry?" An "answer" is based on a formalization of mirror reflections (Chapter 4) using functions. The latter gives a new flavor to geometry and connects it with modern algebra through groups of symmetries. Material in this chapter relates to kaleidoscopes (Chapter 5) and to polyhedra (Chapter 3). Symmetries of frieze and wallpaper patterns, cube and regular tetrahedron get special attention. There are lots of possibilities for exploration with Geometer's Sketchpad.

- **Chapter 7:** The theme of this chapter is "efficient shapes". It's the second chapter with an efficiency theme and considers problems of optimization related to perimeter and area in the plane, surface area and volume in space, and more. This chapter endows geometry with an algebraic cast via the theorem of the arithmetic and geometric mean.

- **Chapter 8:** The chapter's basic question is How to arrange shapes in the plane and space to minimize wasted space? It's the third with an efficiency theme. There are many opportunities to carry out experiments and for making models. Some items of the chapter are at the frontiers of mathematical research; at this writing some questions posed have not been answered.

- **Chapter 9:** This chapter is mainly a big list of project ideas extending themes encountered earlier in book. The projects are for independent research, individually or as part of a group, and may form the basis for final projects in a course using the book. There is an extensive resource list and "How to" section giving suggestions for carrying out a project.

❈ *Success in The Course for Which The Book is Designed*

This is a book for a junior/senior level "topics in geometry" course. Most of the students who take this course will be math majors; many will become high school mathematics teachers. The prerequisites for the course are typically two semesters of calculus and a semester of semi-rigorous linear algebra. Although knowledge of many calculus and linear algebra "facts" are not necessary for success in using the book, ideas from these courses are mentioned; experience with both courses will enhance the experience with this book. The essentials are willingness to dig in and solve a problem (even though at the start you may not have a clue to its solution); a desire to find out whether or not observed patterns persist; interest in mathematical ideas; and eagerness to discover these mathematical ideas on your own. You should be willing to spend time exploring, to take risks, and to learn from mistakes. Persons beginning the book with a view that mathematics is just a "plug and chug" or rote memorization activity will soon have their collective vision shattered. Frustration may be an early experience; excitement and involvement, later ones.

✸ *Use of Technology*

In margins of the text the icon ▦ will occur several times. This indicates that computer software may be useful for visualizing an object, carrying out an exploration, or solving a problem. We will refer specifically to two pieces of software—The Geometry Files and Geometer's Sketchpad®. The Geometry Files is for visualizing three dimensional shapes on the computer screen. The Geometry Files commands enable you in real time to rotate, translate, shrink, expand, and superimpose the shapes. The references to the use of The Geometry Files in the text are of two kinds: use an existing file or create your own file. The separate Instructor's Manual will have information on how to do this. Of course, there are many other pieces of software for visualizing and manipulating three-dimensional shapes, some of which are being developed as I write this. You can purchase The Geometry Files. For information on how to do this, see the next section, The Web.

Geometer's Sketchpad® is a software package for creating and manipulating two-dimensional shapes on a computer screen. With this program you can carry out all the familiar constructions of traditional Euclidean geometry: circle construction, perpendicular segmet bisectors, angle bisection. References in the text to this program are for data-gathering purposes in solving a problem. Of course, this is data gathering of a geometric kind. That's the power of Geometer's Sketchpad. Again, the Instructor's Manual has details on how to use this.

✸ *The Web*

Another use of technology that may be helpful in using this book is the Internet. I will be setting up a Web site for users of this book at http://www.wiley.com/college/gay. Right now it is in its formative, experimental stage, but I plan for you to be able to find there direct links to other Web sites that will enhance the ideas encountered in the book and supplement the book's lists of references. I also hope there will be an opportunity for you to view there what other students have done using this book. I will try to make it possible for you to contribute to the Web site, too. At this site you will be able to obtain more information about ordering The Geometry Files: you will be able to obtain more information there about Geometer's Sketchpad, too.

✸ *Instructor's Manual*

This provides the instructor with summaries highlighting each chapter's important aspects and giving extensive suggestions for spending time in class. It points out problems that are particularly thought-provoking or challenging, ones that are routine, and ones that might be frustrating. It includes solutions to many of the problems and hints for others. The manual contains large-size versions of paper patterns that appear in the text so that the instructor might have copies made on stiff paper for distribution to the students. Finally, the manual includes a number of activities to accompany the text using the computer software *Geometer's Sketchpad* and The Geometry Files.

✳ *Acknowledgments*

I have taught the course for which this book is intended off and on for the last 20 years. During this time I supplemented whatever text I was using with notes. Eventually, the notes became larger than the text and, without my intending it, they became a crude version of this book. Several individuals played a big part in this early development. Students who took the course struggled with the problems and projects and taught me what worked and what got them excited. Their enthusiasm for geometry and for discovery has kept me on the project all these years. I thank them very much. In recent years I have had assistance teaching the course from several of my former students: Ron Hopley, Cathy Miller, Steve Wexler and John Willy. I appreciate very much their help, feedback, and ideas. I also want to express my gratitude to the University of Arizona's Department of Mathematics for supporting curriculum development and valuing innovation in teaching, especially in its teacher preparation program. Special thanks go to my colleagues Fred Stevenson, Larry Grove, and Helmut Groemer for their suggestions and support as I taught the course over the years. I have borrowed freely from several books that have given me mathematical ideas, problems, and approaches. Most of these appear as references in the text. One that deserves special mention is the classic *Induction and Analogy in Mathematics* by Georg Pólya. Its holistic approach to learning mathematics and solving problems has been an inspiration and a guide to me since I was an undergraduate.

I am grateful to Jeanne Deloria for help preparing the manuscript, and to those who reviewed the manuscript, for the many helpful criticisms and encouraging comments during the phases of its development. I would also like to acknowledge support from a National Science Foundation Teacher Preparation grant (to the University of Arizona's Science And Mathematics Education Center) in the form of release time for curriculum development.

Finally, I want to thank my editor, Barbara Holland, and her team at Wiley—especially Ken Santor, who kept us all on task and let me know what was happening at all times; Mario Ferro at EM Technical Art Services, Inc., who created the terrific illustrations; and Maddy Lesure, who designed the cover—for helping to turn those crude notes into something resembling a book. It has been a pleasure and a privilege to work with all of you.

Contents

4 *Shortest Path Problems* 139

5 *Kaleidoscopes* 179

6 *Symmetry* 227

7 *What Shapes Are Best?* 283

8 *Beehives And Other Packing Problems* 327

9 *Where to Go From Here? Project Ideas* 383

Credits 407

Index 409

1 Getting Started: Strategies for Solving Problems

This is a book of problems that are linked together in several ways. They're linked by subject matter: they are problems about points, lines, planes, shapes, ... They are also linked by the very process of solving problems. *Necessity:* To solve one problem you may find it necessary or useful to solve another. *Intrigue:* Solving one problem may spark an interest in solving another. *Connections:* The solution to one problem is really very much like the solution to another. *Vistas:* The solution to one problem (or bunch of problems) may open up a whole new way of looking at things. The linking of these problems forms a dynamic web which, as you begin to know it and see it grow, helps to solve more problems. The object of this chapter is twofold: (1) to throw you smack-dab in the middle of this web; (2) to acquaint you with ways of getting around on the web and maybe help you create some new links. (The web is organic, that is, it's growing. Over time some parts may become more remote and inconsequential than others, while others may become more prominent.) The bottom line is that problems and their solutions create the webbing. What we can provide are a few nodes and a few links. You will fill in the rest. It will be your web. The object now is to get on the web and start "spinning."

This chapter is a "problem solving free-for-all." We'll start with a simple-sounding problem. We'll see, in solving it, that it is connected to other problems. In fact, it will be in dealing with these other problems (thinking about them, solving them) that will lead to a solution of the original problem. You'll see how these connections come about, how you can make these connections yourself. You'll see how the original problem itself actually changes, through clarification, through generalization, through enlarging—and that you will be led naturally to solve a bigger problem than you started with. You will see that your solution is not just barebones (not just a number), but a whole landscape of problems, vistas, arguments, and other paths of connections. Think of this as a prototype for excursions that you will take on your own.

1

✳ *The Problem of Five Planes in Space*

Take five planes in space. Into how many regions do these five planes divide space?

Guesses. Think about the problem a minute. Make a guess for the answer. Keep track of your guess. In fact, take out a blank sheet of paper and mark "Guesses" at the top. From time to time in this discussion as you gain more insight into the problem, make a new guess. Make your first guess now. We'll pause while you do this.

*

What's going on? What is the problem asking for? What does "region" mean? How do you deal with these questions? One way, of course, is just to throw your hands up in horror and say "forget it!" But that's no fun. Another response is to try to think of a simpler, related problem. For example, a simpler problem might be

Into how many regions do four planes divide space?

Well, once you've started this train of thought, how about three planes or even two planes? Imagine two planes. Even here things might not be clear. **How** should I imagine two planes? Take one plane to be the ceiling and another to be the floor. Or rather, these objects extended in all directions indefinitely. Planes are infinite, right? If your room is normal, you can think of these two planes as parallel. From the "side" they look like the following.

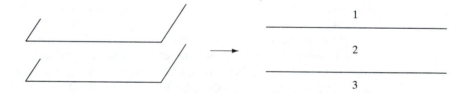

You can see that the two planes divide space up into regions 1, 2, and 3. Does this make things any clearer? We know what "region" means (sort of) and what "dividing space up into regions" means (sort of). But you've probably been thinking "What if the two planes were the floor and one of the walls?"

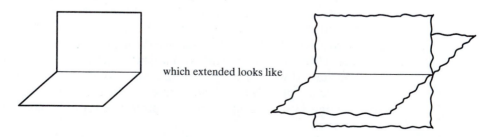

which extended looks like

Looking at these "head on" from the side you'd see the following.

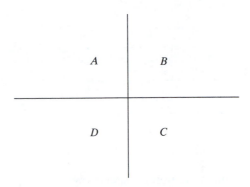

So these two planes divide space up into regions A, B, C, and D. Something's funny. Before, we got three regions. Now we have four. With only two planes, there seem to be two different answers. What might happen with five planes? (Not to forget the original problem.) Should we look for a range of answers? For the moment, let's try for the biggest answer. Later, we can come back and have a look at the other possibilities.

After this clarification our problem is the following.

What is the largest number of regions into which five planes can divide space?

Time to make another guess and write it down on your guess sheet.

* *

We haven't solved the problem yet, but without thinking about it much, we've done something useful. We have a **feel** for the problem. We've **clarified** the problem. How did this happen? **We solved a simpler, similar problem.** We solved the problem for two planes. Since that was such a good idea, let's try the problem with three planes.

Three planes. We know that with two planes different things can happen. Let's look at the possibilities for three planes. What can happen? For two planes the cases were

• two parallel planes
• two intersecting planes

So for three planes we could have

• three parallel planes
• two parallel planes, a third intersecting the other two
• no two planes parallel

The first two cases head-on look like the following

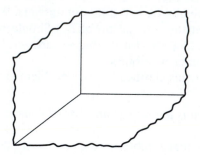

Four and six regions, respectively.

An example of the third case is the floor and two adjacent walls of a rectangular room.

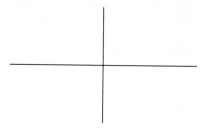

How many regions in this case? One way to visualize the regions is to imagine that one of the planes is the plane of the paper and that the other two are perpendicular to it and make intersections with the plane of the paper as shown below.

So there are four regions formed by the three planes on the other side of the paper and four regions on this side of the paper. Eight regions in all. We've listed all the possible arrangements for three planes. The numbers of regions for the three cases are 4, 6, and 8. So eight is the largest number of regions for three planes.

Let's organize what we have in a chart.

Number of Planes	Number of Regions (largest)
2	4
3	8

There are some even simpler problems that we haven't considered. How about the problem for 1 plane? Or even the problem for 0 planes? These problems sound trivial, but maybe they will give us some insight. We add their solution to our chart.

Number of Planes	Number of Regions (largest)
0	1
1	2
2	4
3	8

Time to add a new guess to your guess sheet. While you're at it, make a guess for four planes, too.

* * *

It looks like we may be getting somewhere. Let's move on to four planes. It appears that things are getting a little complicated for drawing pictures. But it might be a good idea to make a model of four planes, good enough so that all the regions can be counted.

While you're working on that, try thinking of an analogous, simpler problem. We've already dealt with 0, 1, 2, 3 planes in space. This time we're looking for a problem that's simpler, but related in a different way: An analogous problem. Hmm. Planes in space . . . How about lines in a plane? You may have thought about this problem already: You have a bunch of lines in a plane; into how many regions does it divide the plane?

As long as we're thinking about analogous problems, how about one that's even simpler? Yes! Points on a line! You have a bunch of points on a line; into how many regions do the points divide the line?

Let's gather some evidence. Start with 0 points, then 1 point. March right on up . . .

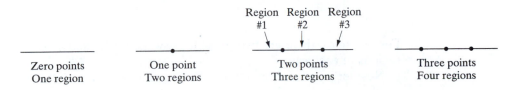

Zero points	One point	Two points	Three points
One region	Two regions	Three regions	Four regions

In fact, if you have n points on a line, then those points divide the line into $n + 1$ regions. We've got a formula! No matter how many points, we can say how many regions! So the number $L(n)$ of regions created by n points on a line is $n + 1$.

$$L(n) = n + 1$$

Here's an argument for this formula. Suppose you have n points on a line. Number the points in order from left to right as they appear on the line.

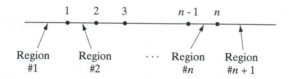

Now let's count regions. There's the infinite region just to the left of point #1. Call that region #1. Then between point 1 and point 2 there is region #2. As you pass each point successively you count another region. Thus between point $n - 1$ and n is region #n and to the right of point n is region #$n + 1$. So $n + 1$ regions in all.

Success! It's time to consider the other analogous problem:

Into how many regions does a bunch of lines divide a plane?

Let's start with a small number of lines and work up, gathering data.

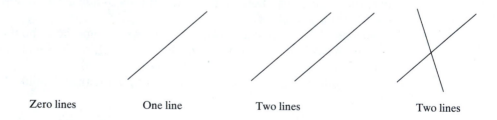

For two lines, there are two possible answers. Let's choose the largest number of regions in each case.

Now let's consider three lines. The possibilities are as follows.

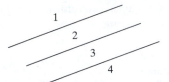

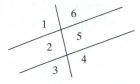

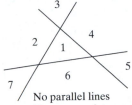

Three lines parallel

Only two lines parallel

No parallel lines
Common intersection

No parallel lines
No common intersection

The number of regions is 4, 6, 6, and 7, respectively.

Problem 1

Are there restrictions on the lines themselves that would guarantee that the number of regions they create is largest? (For example, one restriction might be that no pair of lines be parallel.)

Now for four lines. You're probably thinking: "Let me try to use what I've already done. Start with the three lines on the bottom right above. Add a fourth line, avoiding parallel lines and three or more lines meeting in a point.

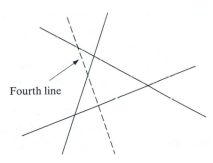

Fourth line

"Let me be careful in counting. With three lines there are seven regions. I want to count new regions created by the fourth line. The total will be the sum of 7 and this new number. How to count the new regions? I notice that the fourth line intersects each of the other three in a point, a different point for each line." (Remember, no parallel lines, no "multiple" intersections?) We can count the new regions as we travel from one "end" of the fourth line to the other. When we reach the first intersection point, a new region has been created. Then, when we reach the second intersection point, another region is added. Similarly, a third region when we reach the third. Finally, there's an additional region created as we complete our trip to the other "end" of the line.

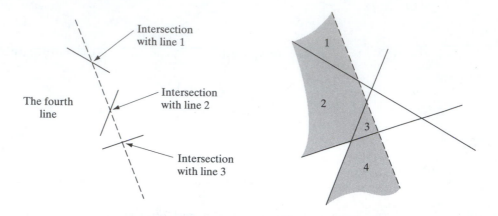

So 7 + 4 = 11 regions in all.

Problem 2

Can you use the counting method we've just described to give you the number of regions created by five lines? six lines? even more lines?

Time to make another guess on your guess sheet. While you're at it, make some guesses for 5, 6, 7, 8 lines.

* * * *

The method we used for four lines can be generalized. Suppose you have n lines and that the number of regions it creates is the largest possible. Call this number $P(n)$. Among these n lines no two are parallel and no three pass through the same point. Now add an $n + 1$st line, making sure that it is not parallel to any of the others and that it doesn't pass through any of the intersections made by the original n lines. So the number $P(n + 1)$ of regions created by the $n + 1$ lines is equal to $P(n)$ + the number of new regions created by the $n + 1$st line. The $n + 1$st line intersects each of the original n lines in a distinct point, a different point for each line. These intersections look like the following.

Each region on this line corresponds to a planar region created by the original n lines and cut in two by the $n + 1$st line. This accounts for all of the new regions. Thus

$$P(n + 1) = P(n) + n + 1$$

Pretty neat! A formula relating the number of regions for $n + 1$ lines with the number for n lines!

Problem 3   Are we really sure that this formula gives us a number which is the largest number of regions created by $n + 1$ lines?

Problem 4   The formula for the function P is called **recursive** because it refers to itself. Can you find a formula for P that is not recursive? (That is, one whose left-hand side is $P(n)$ and whose right-hand side is an expression in terms of n alone and not other values of the function P.)

It's time for a break, time to try some of the strategies used above. Below are problems to get started.

Problem 5   Someone has drawn 1000 points in a plane. No three of them lie on a straight line. You are to connect all pairs of these points with straight line segments. How many of these segments will you have to draw?

Three points   Four points

Problem 6   Five planes all pass through the center of a sphere and divide its surface into regions. How many regions? (If there is more than one answer, find the largest.)

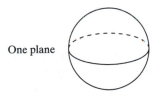

One plane

Problem 7   Ten circles in a plane divide the plane into a bunch of regions. How many regions? (If there is more than one answer, find the largest.)

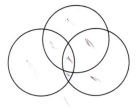

Problem 8   You plan to form an equilateral triangle with 100 pennies on a side. How many pennies will you need?

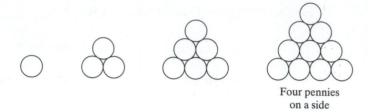

Four pennies
on a side

Problem 9   A triangle made out of pieces of wood is rigid, but a square isn't. It will lean and deform. To prevent a square from deforming, you can add a diagonal bracing. In general, to keep a polygon rigid you may have to add a certain number of diagonal bracings. How many possible diagonals does a 10-gon have? To make it rigid, how many diagonals will you need?

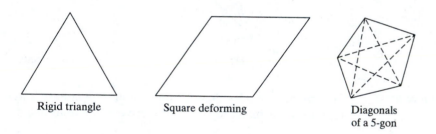

Rigid triangle   Square deforming   Diagonals
of a 5-gon

Problem 10   Ten curved lines are drawn in the plane so that each curved line intersects every other curved line exactly twice. How many intersections are there in all? How many regions are created by the curved lines?

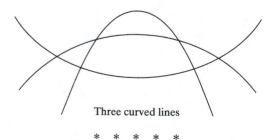

Three curved lines

✳   ✳   ✳   ✳   ✳

It's time to put all we know into a chart. Let's combine what we know about the three situations—regions for points on a line, regions for lines in a plane, and regions for planes in space—into one large chart.

| | Number of Regions Created by | | |
Number of Items	Points on Line	Lines in Plane	Planes in Space
0	1	1	1
1	2	2	2
2	3	4	4
3	4	7	8
4	5	11	?
5	6	?	?
. . .	. . .		
n	$n + 1$		

Notice anything? Have a close look before reading on.

✳ ✳ ✳ ✳ ✳

Wow. Looks like a pattern that works in the following way: To figure out what a number in the third column should be you take the number just above it and add it to the number just above it but in the second column. For example, $11 = 4 + 7$ in the chart below.

| | Number Of Regions Created by | | |
Number Of Items	Points on Line	Lines in Plane	Planes in Space
0	1	1	1
1	2	2	2
2	3	4	4
3	④	⑦	8
4	5	11	?
5	6	?	?
. . .	. . .		
n	$n + 1$		

It looks as if all the numbers we've gotten so far, in columns 3 and 4, can be obtained this way. That's pretty wild if this pattern persists!

If the pattern were to prevail beyond the numbers we've already tabulated, then we ought to be able to answer our original problem. But how do we know if, in fact, it does prevail? Or maybe the pattern doesn't prevail. One way to test the pattern is check it for a particular value. Let's do this for the number of regions created by four planes. If the pattern is valid for all numbers, then the number of regions for four planes in space should be 15.

| | | Number of Regions Created by | |
Number of Items	Points on Line	Lines in Plane	Planes in Space
0	1	1	1
1	2	2	2
2	3	4	4
3	4	7	8
4	5	11	15?
5	6	?	?
. . .	. . .		
n	$n + 1$		

The "?" after the 15 indicates our uncertainty as to whether or not 15 is right. How can we decide? (If 15 is not the right number, then we'll have to toss our pattern out the window . . .) The other numbers in the table were obtained either by drawing a picture and counting regions or, as in the case of the later numbers in column 2, by consequence of an argument. At this stage our best bet is to count regions. Here is where we need the model you made for this situation. Your model might look something like the object shown in the following picture.

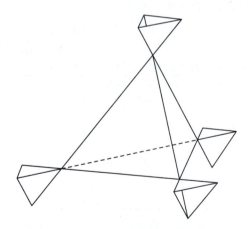

Even with a model the situation is complex. How do you know if you have counted everything? Maybe the analogous problem (lines in the plane) might give us a clue. What makes the case of four planes unusually difficult? It's the first time a bunch of planes has created a bounded region. For lines in plane, this occurs first with three lines.

Let's count regions for three lines again, this time being a bit more systematic. Maybe classifying the regions would help. The bounded region is a triangle. There's a region at each vertex of the triangle for a total of three (3). (See the speckled parts of the diagram below.) There's a region at each edge for a total of three (3). (See the cross-hatched part of the diagram below.) And, of course, the interior of the triangle itself for one (1) more. The total number of regions is $3 + 3 + 1 = 7$.

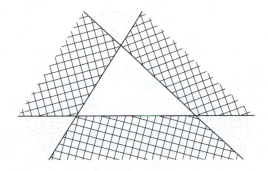

In the case of four planes, the bounded region is a *tetrahedron*, a shape in space with four triangular faces and a figure in space analogous to a triangle in the plane.

The ingredients for a triangle which helped us classify regions in the plane were vertices, edges, and interior. The analogous ingredients for a tetrahedron are its vertices, edges, faces, and interior: One region for each vertex (4); one region for each edge (6); one region for each face (4); and the interior of the tetrahedron itself for one (1) more. Total number of regions is $4 + 6 + 4 + 1 = 15$. This should jibe with what you counted on your model and the picture above. It certainly jibes with our prediction!

Problem 11   How do you know that 15 is the largest number of regions created by 4 planes? (see also Problem 3.)

What do you think now about the validity of the pattern for all numbers of planes and all numbers of lines? If the pattern holds for five planes, then the number of regions created by five planes is 26.

| | | *Number of Regions Created by* | |
Number of Items	*Points on Line*	*Lines in Plane*	*Planes in Space*
0	1	1	1
1	2	2	2
2	3	4	4
3	4	7	8
4	5	11	15
5	6	?	26?
. . .	. . .		
n	$n + 1$		

How to decide? We could build a model for five planes and count—too much work. Also, at this point we're committed to more. I think we'd like the whole enchilada. We think we know what's going on for all planes. Maybe we should go for a general argument, a proof.

Before doing that, let's spell out in another way what we think is going on. Let $S(n)$ denote the largest number of regions created by n planes in space. Recalling that $P(n)$ denotes the largest number of regions created by n lines in a plane, our claim for planes in space is the following:

$$S(n) = S(n - 1) + P(n - 1)$$

This looks vaguely familiar. Remember we showed that

$$P(n) = P(n - 1) + n$$

If we let $L(n)$ denote the number of regions created by n points on a line, then this latter formula becomes

$$P(n) = P(n - 1) + L(n - 1)$$

This is just a restatement of the pattern we observed for column 3 in functional notation! Moreover, we've shown this to be true for all n. This suggests even more strongly that we are on the right track. Incidentally, with this notation and insight our chart becomes the following!

n	$L(n)$	$P(n)$	$S(n)$
0	1	1	1
1	2	2	2
2	3	4	4
3	4	7	8
4	5	11	15
5	6	16	26?
. . .	. . .		
n	$n + 1$		

Note. $L(n)$ = number of regions on a *Line*; $P(n)$ = number of regions on a *Plane*; $S(n)$ = number of regions in *Space*.

It looks like what we have shown for regions created by lines on a plane general-izes to (can also be shown to occur for) regions created by planes in space.

What is the status of the formula $S(n) = S(n - 1) + P(n - 1)$? We know that the formula is true when $n = 0, 1, 2, 3,$ and 4. Our **conjecture** is that

$$S(n) = S(n - 1) + P(n - 1) \text{ is true for all } n$$

The truth of the conjecture is another way of saying that the pattern prevails. We have a strong feeling that this statement is true, and we have a lot of evidence to back it up.

Incidentally, another conjecture suggested by "early" evidence is

$$S(n) = 2^n \text{ is true for all } n$$

My guess is that you might have come up with this fairly early in the discussion. The formula $S(n) = 2^n$ is true for $n = 0, 1, 2, 3$. Is it true for all n? No. In fact, $S(4) = 15 \neq 2^4$. The latter is called a **counterexample** to the conjecture. A counterexample proves a conjecture false, as it does in this case.

We will leave the resolution of truth of the conjecture "$S(n) = S(n - 1) + P(n - 1)$ is true for all n" to you in the following problem.

Problem 12   If you were going to find an argument proving the conjecture $S(n) = S(n - 1) + P(n - 1)$ is true for all $n \geq 1$ (where $S(0) = P(0) = 1$), a good tactic might be to try an argument analogous to the one we used to prove $P(n) = P(n - 1) + L(n - 1)$ is true for all $n \geq 1$ ($P(0) = L(0) = 1$). Try it. (Of course, you might also try to find a counterexample. A counterexample would prove the conjecture to be false.)

Problem 13   Suppose you have a function $F(n)$ such that $F(n) = F(n - 1) + P(n - 1)$ for all $n \geq 1$ and $F(0) = 1$. [The function $P(n)$ is defined as above.] The formula for $F(n)$ is recursive. Find an alternative formula for $F(n)$ which is not recursive. You might want to use the nonrecursive formula you obtained for $P(n)$ earlier.

Problem 14   If you have five planes in space, the number of regions they create may depend on how the planes are arranged. We decided to narrow the original problem and find the maximum possible number of regions created by five planes in space. Go back and see if you can fill in what we blanked out: what are the possible numbers of regions that five planes could create?

✳ *Strategies for Solving Problems*

Let's have a look at some of the strategies we used to solve the Five Planes in Space Problem.

Make a Guess. This is not a silly device. It enabled us to get started and to keep going. The idea is not to make a random guess, but a guess based on the most informed view of the problem you have at the moment. The view might be erroneous, but that's o.k. The idea is to keep revising your guess as you get more experienced with different aspects of the problem. You may not always be able to articulate exactly why you have made a certain guess. Articulating it is part of the process, too. Having a list of guesses that you revise from time to time will also help you to keep on track and to measure your progress on the original problem.

Solve a Simpler, Related Problem. This is a way of getting insight on the original problem. Hard to visualize a problem? Solve a simpler one that you can visualize. Also, if you're stuck, it's a way to keep you thinking about the problem, a way to keep you from giving up.

Clarify the Problem. After digging around a bit, it seemed as if our problem could have several answers. We made a choice and decided to look for the largest answer. We narrowed the focus of the problem. Once you solve the narrower problem you might want to go back and see if there's more to do in solving the original problem.

Solve an Analogous Problem. This strategy is similar to "Solve a simpler problem." You choose to solve an analogous problem because you understand the analogous problem better, or because it seems simpler, or because you feel more comfortable with it. For us, the solution to the analogous problem gave us insight on the original problem. A formula for the analogous problem suggested a formula for the original problem. Detailed analysis of one case of the analogous problem suggested a way to carry out a detailed analysis for one case of the original problem. A successful argument for the analogous case may suggest a successful argument for the original problem. Solving the analogous problem also gave us a feeling that we were getting somewhere. It energized us. Even if what worked with the analogous problem doesn't work with the original problem at least it gives us a kind of benchmark for comparison. "Hey! It worked here, but doesn't work there. That's interesting! Why doesn't it work there?" It suggests questions whose answers might lead us to a solution. It keeps us going. In the words of that famous detective, Hercule Poirot, "It keeps those little grey cells working."

Draw a Picture; Make a Model. If the problem suggests a picture or diagram, then sketch it. Or draw a picture for a simpler problem or for an analogous problem. In our case, the simpler problems and the analogous problems might have popped into our heads simply because we could draw pictures for them! A good picture brings your visual apparatus into your problem-solving arsenal. Looking at the picture—elaborating it, adding labels to it—may suggest something that wouldn't have occurred to you otherwise. This is a good strategy for all problems. It is absolutely essential for geometry problems! A model is just an elaborate picture. You may learn something just by building the model. Examining a model brings your tactile senses into your problem-solving arsenal. If it's a three-dimensional model, then using it will enable you to bring your spatial senses into solving the problem. In our problem, a model was useful in verifying an instance of a conjecture. A model (especially when you make it yourself) has a way of getting you hooked: its tactile qualities, its beauty, . . . make you want to solve the problem like you never did before.

Organize Data in a Chart. In the course of solving a problem, try to organize your solutions to simpler (or analogous) problems in a chart or table (these solutions are your data), especially if the problems can be ordered by degree of complexity. A chart is another visual device to give you insight into your problem and to suggest side problems for further investigation.

Look for Patterns. Once you have your data organized, look to see if you can find a pattern (or trend). The trend we observed in our chart was that, as we moved down column 4, the number there was equal to the number just above it (in column 4) plus the number above it and to its left (in column 3). There are other patterns that can be observed. For example, we might have observed that as you move down column 3, the differences between the values form the sequence 1, 2, 3, 4, ...

Make a Conjecture. Then, once you have observed a pattern, it's time to commit yourself. A conjecture states that the pattern you observe is true not just for the cases you have observed but for all possible cases. (If your conjecture does not have the solution to your original problem as one of its consequences, eventually you will want to make a conjecture that does.) Making a conjecture out of several observed instances is sometimes called generalizing (the observed instances).

Verify Consequences. Having made a conjecture, you should try to verify that one or more consequences of the conjecture are true, consequences that you don't know are true before stating the conjecture. For example, the conjecture we made above stated $S(n) = S(n-1) + P(n-1)$ for all $n \geq 1$ (with $P(0) = S(0) = 1$). Some instances of this conjecture we already knew to be true: the truth of the formula for $n = 1, 2$, and 3. What we didn't know was its truth for $n = 4$. We then verified the formula for this case. Since the game is to show the formula true for all $n \geq 1$ (see the next strategy "Prove the Conjecture. . ."), this gives us more confidence in our conjecture. For, if we were to find out that the formula were not true for $n = 4$, then we would have to throw out the conjecture and look for another pattern that explains the data.

Prove the Conjecture or Find a Counterexample. We never know whether a conjecture is true until we find an argument proving it to be so. Knowing that several instances of the conjecture are true is not enough. In the preceding discussion, we saw a conjecture with several instances of it true, yet the conjecture itself was false. Although the evidence in favor may seem overwhelming and our commitment to it very strong (because we observed the pattern, or for whatever reason), we still don't know that the conjecture is true until we have an argument that justifies it.

Suppose that the solution to a problem we're interested in is an instance (or, more generally, a consequence) of the conjecture. There are several reasons why providing a proof of the conjecture is a good idea. First, proving the conjecture would solve the problem. Second, we get some insight into the pattern we have observed. We learn thereby that the pattern we have observed has an explanation, that the individual instances are not just coincidental, but hang together by their connection with other things we already know to be true. Of course, it may be that the conjecture is not true. All we need in order to know that it is not true is to exhibit a counterexample—an instance of the conjecture that is not true. All you need is one false instance to find the conjecture false. If you do find a counterexample, you may not just want to throw out the baby with the bathwater. You may want to revise the conjecture, taking into account the information carried by the counterexample.

How you come up with an argument proving a conjecture is something else. But finding an argument is a lot like the problem-solving process we have been describing here. Find conjectures for similar, simpler situations and construct arguments (or find counterexamples) for them. The arguments you use in these simpler settings may give you ideas for the conjecture you're really interested in. Maybe an argument you've used for another situation can be adapted to this one. One thing is certain, however: the more experience you have finding arguments, the more successful you will be. The latter is another way in which finding arguments is like solving problems: the more proofs you construct, the better "prover" you will be. I guarantee it.

✳ *Baby Gauss and a Formula*

An expression you may have encountered in trying to come up with formulas for some of the earlier problems is the following.

$$1 + 2 + \cdots + n$$

i.e., the sum of all the numbers from 1 to n.

From the way it is written you need to make $n - 1$ additions in order to calculate the value of the expression. The number of calculations increases as n gets bigger. There is another expression (involving n) equal to this which for every n involves two operations, a multiplication, and a division, in order to calculate its value. Legend has it that this latter expression is due to Carl Friedrich Gauss when he was a schoolchild.

Gauss (1777–1855) was one of the greatest mathematicians of all times. The story goes that when Gauss was in an early grade at school his classmates were being particularly rambunctious. In order to settle them down the teacher devised a task for them to solve at their desks. The task was to add up all the numbers from 1 to 100. The teacher, who thought this would occupy them for a while, was surprised when a few moments later Gauss raised his hand with the answer.

"How did you get it so fast?" the teacher asked.

"Easy," said Gauss. "Imagine that you had written all the numbers down on paper like this to add them up:

$$
\begin{array}{c}
1 \\
2 \\
3 \\
\vdots \\
98 \\
99 \\
100
\end{array}
$$

Then beside these numbers you wrote them again, only in the opposite order:

$$
\begin{array}{cc}
1 & 100 \\
2 & 99 \\
3 & 98 \\
\vdots & \vdots \\
98 & 3 \\
99 & 2 \\
100 & 1
\end{array}
$$

Let's add up all of these numbers. I'll get twice as much as I want but I can divide by two when I get done. But, instead of adding up and down twice, I'll do all the little additions sideways:

$$
\begin{array}{ccccccc}
1 & + & 100 & = & 101 \\
2 & + & 99 & = & 101 \\
3 & + & 98 & = & 101 \\
\vdots & \vdots & & & \\
98 & + & 3 & = & 101 \\
99 & + & 2 & = & 101 \\
100 & + & 1 & = & 101
\end{array}
$$

There are 100 little additions, all with the same answer! So the sum of the two columns of numbers on the left is the same as the single column of numbers on the right. The latter is $100 \times 101 = 10100$. Half of that is 5050. And that's the answer!"

Amazing!

Here is Gauss's argument in general, with vertical and horizontal switched:

$$2(1 + 2 + 3 + \cdots + n)$$

$$
\begin{array}{llllllllllll}
= & 1 & + & 2 & + & 3 & + \cdots + & n-2 & + & n-1 & + & n \\
+ & n & + & n-1 & + & n-2 & + \cdots + & 3 & + & 2 & + & 1
\end{array}\left.\right\}\begin{array}{l} \text{add} \\ \text{vertically} \end{array}
$$

$$= \underbrace{n+1 \ + \ n+1 \ + \ n+1 \ + \cdots + \ n+1 \ + \ n+1 \ + \ n+1}_{n \text{ times}}$$

$$= n(n + 1)$$

Dividing both sides by 2, we get

$$1 + 2 + \cdots + n = \frac{n(n + 1)}{2}$$

✳ A Collection of Problems

Here are some problems to have fun with. Some are closely related to the problems you have seen earlier, and some are new. Many of the problems are related among themselves. So if you are having trouble with one problem, you may want to look at another to see if what you might do **there** would help you **here**. (You've heard this before!) You might even find it more fun to stay **there**!

The problems are grouped loosely by themes. You will encounter some of these themes (or variations on them) later in this book.

Try out your arsenal of problem-solving skills. You will find that you won't need them all for each problem. Or at least you won't need them in the same order or to the same degree. Things won't work quite the same way as they did in the five-planes problem. After all, working on a problem is an adventure. If you knew what was ahead, it wouldn't be one. You wouldn't have surprises, and it wouldn't be fun! Dullsville.

Success begets success. As you do more problems, you'll get better at it. Not by starting over again each time, but by building on what you have done. To help you in the building process, it would be a good idea to document what you do. Do this in two ways. One, write up each solution carefully and thoroughly in such a way that the solution can be reconstructed later by you and also by somebody else—the latter in order to get some feedback from someone else. Maybe you got off on a wrong track, maybe you missed an important part, maybe it's just wrong (!) . . . and it takes someone not as engrossed in the problem as you were to see this and point it out. In any case, solving problems is too much fun to keep to yourself. Tell your solutions to your friends and teachers. They can be supportive and also can give you some ideas which might improve your understanding or which might help you clean up a messy argument.

Two, record your solutions—in a notebook or folder—so you can refer to them later for help in solving other problems and so you can keep track of your progress as a problem solver.

LATTICE (GEOBOARD) PROBLEMS

Loosely, a lattice is a neatly arranged array of dots covering the plane. One lattice is called the **triangular lattice**. A piece of this is shown below. Three connected dots in close proximity form a **unit triangle**, which also happens to be equilateral.

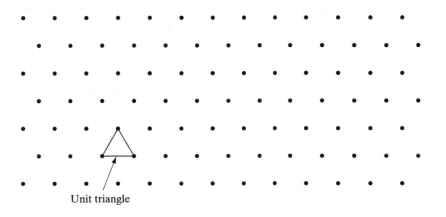

Unit triangle

Problem 15   Form a large equilateral triangle 100 (horizontal) layers tall by connecting dots. By connecting dots, you can form a lot of unit triangles inside the large one. How many?

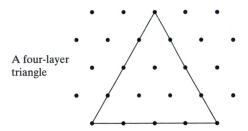

A four-layer triangle

Problem 16   On the triangular lattice form a large equilateral triangle 100 layers tall (as in problem 15). By connecting dots in this triangle you can find parallelograms of various sizes, each side of which is parallel to a side of the big triangle. How many sizes of parallelograms can you form? How many parallelograms of all sizes can you form in the large triangle?

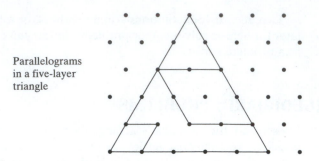

Parallelograms
in a five-layer
triangle

Another lattice is the square lattice, a piece of which is shown below. Connect four dots in close proximity and you get a unit square.

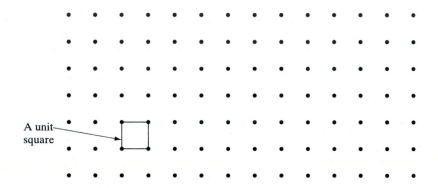

A unit square

Problem 17   Connect the dots to form a large square 100 layers tall. In this large square you can connect dots to form a lot of little unit squares. How many? You can also form squares of other sizes in the large square. (Assume the bases of these squares are horizontal, that is, parallel to bottom of page.) What are the sizes you can form? How many squares of all sizes (horizontal bases) can you form?

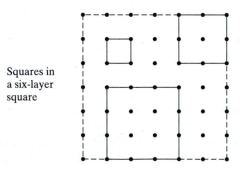

Squares in
a six-layer
square

Problem 18 Connect the dots on a square lattice to form a rectangle 50 layers tall and 100 layers wide. How many squares of all sizes with horizontal bases can you form in the large rectangle?

Problem 19 In a large square 100 layers tall, form rectangles with horizontal bases. How many different sizes can you form? How many different rectangles of all sizes?

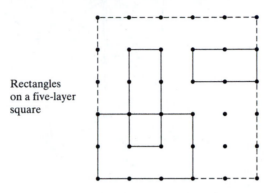

Rectangles
on a five-layer
square

Problem 20 In a rectangle n layers tall and m layers wide, form rectangles with horizontal bases. How many different sizes can you form? How many different rectangles of all sizes?

Problem 21 In the square lattice in the plane, assume that the unit square has area 1. In a square 100 layers deep you can create lots of triangles by connecting dots. The object of this problem is to investigate the areas and perimeters of these triangles. Here are some questions.

- Try to create triangles having the following areas: 1, 2, 3, 40, 41, 1/2, 3/2, 1/3, 3/10, 3/4. What happens?
- How many different (i.e., noncongruent) triangles can you form of area 1? (What are they?)
- Of all triangles having area 1, which one has maximum perimeter? Which one has minimum perimeter?
- Of all triangles (any area) that can be formed in the square, which ones have the largest perimeter? Smallest perimeter? Largest area? Smallest area?

BOX PROBLEMS

Problem 22 In space there is an array of dots called the **cubic lattice**. Connect eight dots in close proximity and you get a **unit cube**.

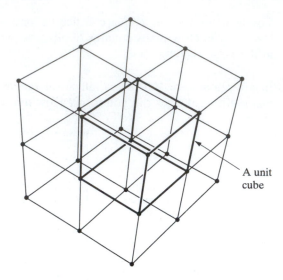

A unit
cube

Connect dots in the cubic lattice to form a cube 100 layers high. In this large
cube you can connect dots to form a lot of little unit cubes. How many? You
can also form cubes of other sizes in the large cube. (Assume the bases of these
cubes are horizontal.) What are the sizes you can form? How many such cubes
of all sizes can you form?

Problem 23   Problems 18, 19, and 20 suggest possible generalizations of the previous problem.
State them and solve them.

PYRAMID PROBLEMS

Problem 24   In problem 8, the number of pennies in the figure formed by 100 layers of pennies
is called the 100th *triangular number*. In this problem, we will "stack" triangles
to form pyramids. To do this replace each penny in an *n*-layered figure by a
marble of the same radius. Form a 100-layered pyramid by placing a 99-layered
triangle on top of a 100-layered triangle, then a 98-layered figure on top of the
99-layered triangle, and so on, with a 1-layered triangle on the very top. The
final shape is a triangular pyramid, each face of which is a copy of the 100-layer
triangle of problem 8. (Get some marbles, styrofoam balls, or something, and
make a smaller version of this!) The number of spheres making this pyramid is
the 100th *pyramidal number*. What is this number?

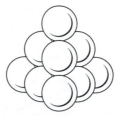

A two-layer pyramid   A three-layer pyramid

Problem 25   Here is another way to build a pyramid of spheres. Take a 100 by 100 square array of spheres. On top of that place a 99 × 99 square array of spheres, then a 98 × 98 array on top of that, and finally a 1 × 1 array of spheres on the tippy top. How many spheres make up this pyramid?

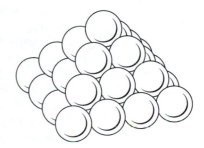

HANDSHAKE PROBLEMS

Problem 26   At a large party of 100 people, each person shakes hands with everybody else just once. How many handshakes occur?

Problem 27   You throw a party for 10 husband-and-wife couples. (You and your spouse form one of these couples.) During the party several people shake hands. (Of course, a husband does not shake hands with his wife, and **vice versa**). At the end of the party you ask each person how many individuals he or she shook hands with. The responses are all different. How many persons did your spouse shake hands with?

PATH PROBLEMS

Problem 28   A worker bee, in the lower left cell of this strange bee hive, wants to visit the Queen Bee located in some other cell. To go there, the worker bee must move to one cell at a time. From an upper cell he can move either to its neighbor directly to its right or to its neighbor down and to its right; from a lower cell he can move either to its neighbor to its right or to its neighbor up and to its right. If the Queen is in cell 50, how many paths can the worker take getting there?

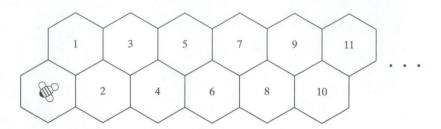

Problem 29

Exploratory Problems in Mathematics, by Fred Stevenson (1993), p. 68, contains the following problem.

I have accepted a job in the city. I requested that my employer find me a home less than one mile from my workplace so that I could bicycle to work every day. I also wanted my home to be placed so that my trips to work could be along a different route for every day of my contract. My contract is for five years.

I will work 5 days a week, 50 weeks a year for 5 years. The streets in this area of the city form a grid of squares, each 1/16 mile on a side; therefore, my home must be fewer than 16 blocks from work.

At how many different sites could my home be located?

What is the nearest that I can live to my place of work and still have access to enough different routes?

At what sites could I get the maximum number of routes to work? How many years could I work at this job and ride to work along a different route each day if my home is located at a site of maximum routes?

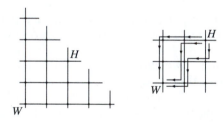

TILING PROBLEMS

Problem 30

Take a lot of squares of the same size.

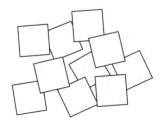

You can arrange these squares as shown.

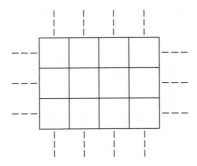

If you had enough of these squares, you could extend this arrangement and tile the entire plane; that is, you could cover the plane completely with these squares with no overlaps. It's like tiling your kitchen with square linoleum or ceramic tiles, except that the plane goes on forever in all directions; it needs infinitely many tiles for a complete covering. This covering of the plane by many copies of a single shape is called a **tessellation.** Can you tile the plane with equilateral triangles? Can you tile the plane with pentagons, each with all its sides equal and all its interior angles equal? The tiling with squares has two squares meeting full edge to full edge. You want a tiling with triangles or pentagons to satisfy this property, too.

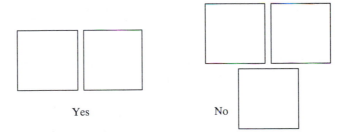

Recall that a regular *n*-gon is a polygon with *n* sides having all its sides equal and all its interior angles equal. Thus an equilateral triangle is a regular 3-gon; a square is a regular 4-gon. For which *n* can you tile the plane with regular *n*-gons?

Problem 31 A number of triangles are shown below. Make several copies of each of them. Try to make a tiling with each of the triangles. Which of the triangles will tile the plane? What can you say in general about a triangle that will tile the plane?

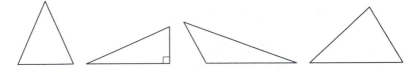

Problem 32 A number of quadrilaterals are shown. Make several copies of each. Try to make a tiling with each of the shapes. Which of the quadrilaterals will tile the plane? What can you say in general about a quadrilateral that can tile the plane?

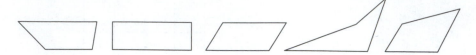

✳ *A "Trick" Yields Another Formula*

Earlier we talked about Gauss's method for finding a "closed" expression for the sum of the first n numbers.

$$1 + 2 + 3 + \cdots + n = \sum_{k=1}^{n} k = \frac{n(n+1)}{2}$$

The following is a method that does the same for the sum S of the first n squares:

$$S = 1^2 + 2^2 + 3^2 + \cdots + n^2$$

The method involves an algebraic maneuver you might not think of using right off the bat. It involves creating an expression for n^3 that is a summation of n terms. This elaborate expression is what is called a **telescoping series**—"adjacent" terms of the expression cancel each other and make most of the expression "collapse." Here it is:

$$n^3 = n^3 - (n-1)^3 + (n-1)^3 - (n-2)^3 + (n-2)^3 - (n-3)^3 + \cdots + 3^3 - 2^3$$
$$+ 2^3 - 1^3 + 1^3 - 0^3$$

$$= [n^3 - (n-1)^3] + [(n-1)^3 - (n-2)^3] + [(n-2)^3 - (n-3)^3] + \cdots + [3^3 - 2^3]$$
$$+ [2^3 - 1^3] + [1^3 - 0^3]$$

$$= \sum_{k=1}^{n} [k^3 - (k-1)^3]$$

$$= \sum_{k=1}^{n} [k^2 + k(k-1) + (k-1)^2] \ \{\text{since } a^3 - b^3 = (a-b)(a^2 + ab + b^2)\}$$

$$= \sum_{k=1}^{n} [k^2 + (k^2 - k) + (k^2 - 2k + 1)]$$

$$= \sum_{k=1}^{n} [3k^2 - 3k + 1]$$

$$= 3 \sum_{k=1}^{n} k^2 - 3 \sum_{k=1}^{n} k + \sum_{k=1}^{n} 1$$

$$= 3S - 3 \left[\frac{n(n+1)}{2} \right] + n$$

In other words

$$n^3 = 3(1^2 + 2^2 + \cdots + n^2) - \frac{3n(n+1)}{2} + n$$

Or, solving for $1^2 + 2^2 + \cdots + n^2$, we get

$$\sum_{k=1}^{n} k^2 = 1^2 + 2^2 + \cdots + n^2 = \left[n^3 + \frac{3n(n+1)}{2} - n \right] \Big/ 3$$

$$= \frac{n(n+1)(2n+1)}{6} \qquad \{\text{using a little algebra on the right-hand side}\}$$

Neat, eh?

Problem 33   Using the method for the sum of the first n squares, find a closed expression for the sum of the first n cubes.

✳ *Proofs Without Words*

Sometimes pictures can be used to "prove" a formula. For example, here is a visual argument for the formula for the sum of the first n numbers. Use dots to represent the numbers and arrange the dots in a right triangular formation.

Make a copy of this triangle with lighter colored dots; rotate the latter 180 degrees and place the two triangles together to make a rectangular array of dots with n rows and $n + 1$ columns.

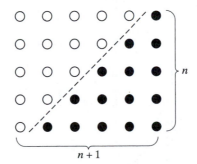

The rectangular array contains $n(n + 1)$ dots. Half the array contains $n(n + 1)/2$ dots. And half the array is also the triangle made up of $1 + 2 + 3 + \cdots + n$ dots. Thus $1 + 2 + 3 + \cdots + n = n(n + 1)/2$. Thus the last picture above in some sense encapsulates a proof of the formula.

Problem 34   A number of pictures are shown below. What formulas do they suggest? Turn them into proofs of the formulas.

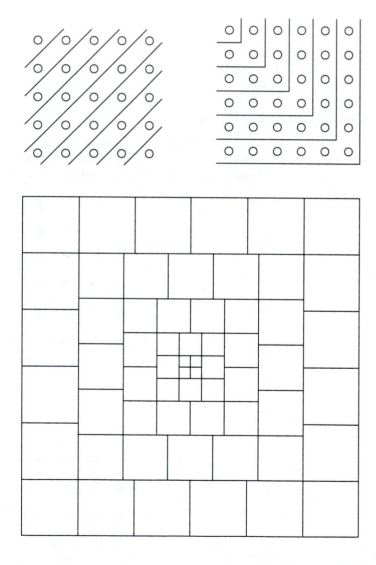

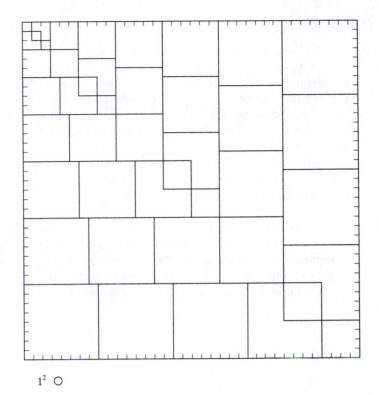

1^2 ○

2^2

3^2

4^2

5^2

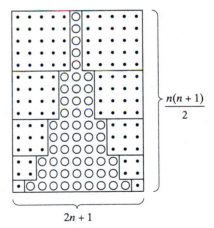

$$\frac{n(n+1)}{2}$$

$$2n + 1$$

✳ *Mini-Projects*

The following problems are a little bit more open-ended and less routine than what you have seen. So we call them mini-projects. Some appear to be extensions of problems that have appeared earlier; others are new. Approach these problems as you would the other problems you and I have been working on: make guesses, gather and organize data, look for patterns, make conjectures, look for counterexamples, prove your conjectures, ... Expect to spend a little time finding a solution and writing it up.

Your class might choose to tackle these projects in such a way that a team of three or four students works on a single project. Guidelines for how this might be done are given at the end of this section.

Mini-Project 1   In a 100-layer equilateral triangle formed on a triangular lattice, you can form a lot of unit equilateral triangles (see problem 15). You can also form a lot of equilateral triangles of other sizes, some right-side up, some up-side down, but all with one side parallel to the large triangle's base. How many different sizes? How many right-side up and up-side down of all sizes?

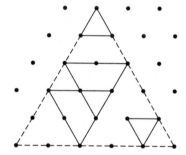

Triangles in a five-layer equilateral triangle

Mini-Project 2   In a 100-layer equilateral triangle formed on the triangular lattice, you can also form smaller equilateral triangles whose bases do not lie along the horizontal. How many sizes of such equilateral triangles of any orientation can you form? How many such equilateral triangles of all sizes and orientations can you form?

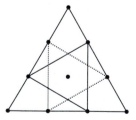

Triangles in a three-layer equilateral triangle

Mini-Project 3   Consider a 100-layer equilateral triangle formed on the triangular lattice. In problem 16 you formed parallelograms each of whose sides was parallel to a side of the triangle. This time, form parallelograms of all orientations and sizes. How many different orientations and sizes can you form? How many parallelograms of all sizes and orientations can you form?

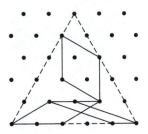

Mini-Project 4   In a 100-layer square formed on the square lattice in the plane, you can form smaller squares whose bases do not lie along the horizontal. How many sizes of squares of any orientation can you form? How many squares of all sizes and orientations can you form?

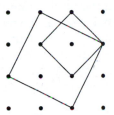

Mini-Project 5   You can think of the points in the cubic lattice as being all points in space having integer coordinates. Thus in the cubic lattice there are three "preferred" directions: the x-axis, the y-axis, and the z-axis. And there are three preferred coordinate planes: the x-y plane, the x-z plane, and the y-z plane. In a 100-layer cube formed on the cubic lattice with sides parallel to the coordinate planes, you might be able to connect dots and form cubes such that one or more of its sides is not parallel to the three coordinate planes. What are the possibilities?

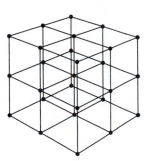

Mini-Project 6 Take 10 points on a circle. Connect all pairs of these points to form chords of the circle. Choose the points on the circle so that no three chords intersect in a single point interior to the circle. The chords divide the circle into how many regions?

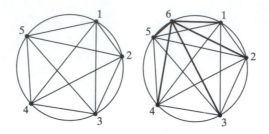

Mini-Project 7 Problem 30 deals with the tiling of the plane by copies of a single regular polygon. For example, you can tile the plane using a bunch of squares of the same size. You have become bored with making a tiling pattern with just one shape. You want to use several differently shaped tiles in your tiling pattern. You still want them to cover the plane without overlap and without gaps and you still want to use regular polygons. But you will allow triangles **and** squares; or triangles, pentagons, and octagons. You want the polygons to fit together edge-to-edge so you want the lengths of all the polygons to be the same; you want the configuration of polygons at any one vertex to be the same as at any other vertex. You want to know what the possibilities are.

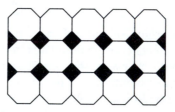

Mini-Project 8 In problem 21 we formed triangles in a 100-layer deep square drawn on the square lattice in the plane. This project is an investigation of the area, perimeter, number of interior lattice points, and number of boundary lattice points of such triangles and of other polygons formed by connecting dots in the square lattice. The following are questions to get you started.

- What is the largest area of a triangle having no interior pegs?
- What is the smallest area of a triangle having no interior pegs?
- What is the largest perimeter of a triangle having no interior pegs?
- What is the largest number of boundary pegs on a triangle?

- What is the largest area of a triangle whose only boundary pegs are its vertices?
- What is the largest perimeter for such a triangle?
- Can you find a relationship among the number of boundary pegs, the number of interior pegs, and the area of a triangle?
- How many different polygons of area 2 can you draw on the square lattice?
- Does the relationship for triangles extend to polygons drawn on the square lattice?
- Can you find a similar relationship among the volume, interior dots, boundary dots, etc. for triangular pyramids (and other polyhedra) formed by connecting dots on the cubic lattice in space?

Mini-Project 9   WYTHOFF's game. Think of the square lattice in the plane as all points having integer coordinates. In this two-person game the playing board consists of all the integer lattice points in the upper right quadrant, that is, all (n, m) such that $n, m \geq 0$. A dime is placed on the lattice point $(29, 47)$. The two players take turns moving the dime to another point on the playing board. At each turn the player may move the dime as far as desired on the playing board but it must be moved in one of the following directions:

- horizontal, from right to left ←
- vertical, from top to bottom ↓
- diagonal, from upper right to lower left (parallel to the line $y = x$)

The player to reach the origin $(0, 0)$ first is the winner.
Would you like to be the player who makes the first move?

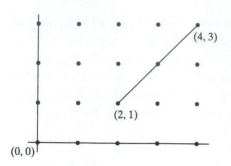

Mini-Project 10   (a) An $n \times m$ rectangle (n, m positive integers) is partitioned into nm equal squares (in the obvious manner). A diagonal of the rectangle passes through many of these squares. How many?

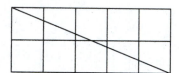

(b) Generalize part (a) to an $n \times m \times p$ rectangular box and one of its diagonals.

Mini-Project 11   Arrange three pennies to form an equilateral triangle. By moving just one penny you can turn the arrangement "upside down."

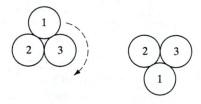

Now make an arrangement of pennies that forms an equilateral triangle with 1000 pennies on a side. (See the figures in problem 8.) How many pennies do you have to move to turn this arrangement upside down?

Mini-Project 12   A ball hit by a cue on a billiard table is assumed to behave in the following fashion when it hits a side.

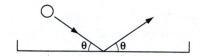

Suppose that the playing surface of a billiard table is an $n \times m$ rectangle (n and m positive integers). Suppose you shoot a ball out of one corner at a 45° angle to one of the sides. What can you say about the path of the ball?

Assume that if a ball hits a corner, then it falls into a "pocket" and stops. For the 1×1 table, the ball eventually goes into the pocket opposite from where it was hit. Does this happen with other $n \times m$ tables? What is the length of the path of the ball before it falls into a pocket? How many rebounds does the ball make off the sides of the table before it falls into a pocket?

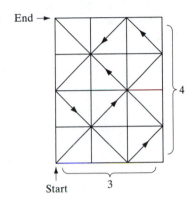

Mini-Project 13 A carpenter, working with a buzz saw, plans to cut a wooden block 12 in. by 32 in. by 96 in. into $12 \times 32 \times 96 = 36{,}864$ little 1-in. cubes. After some cuts have been made, he can make a single cut by sawing through a stack of pieces. He wants to accomplish the job by making the fewest number of cuts. How? And how many cuts will he need?

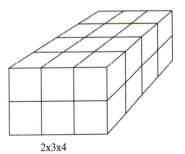

2x3x4

Mini-Project 14 [Adapted from *Exploratory Problems in Mathematics*, by Fred Stevenson (1993).] How many triangles with integer sides have perimeter equal to 1,000,000? How many of these are equilateral? isosceles? scalene?

Turn the question around. What would the perimeter have to be if you wanted to be able to form at least 1,000,000 triangles? isosceles triangles? scalene triangles?

Mini-Project 15   [Adapted from *Exploratory Problems in Mathematics* (1993).] In problem 29 we lived in an area of a city where the streets form a grid of squares. Our friends the prairie dogs live underground in neighborhoods like this, with streets going in the perpendicular directions of north-south and east-west, but they also build in the up-down direction. Imagine that the job situation is exactly the same as in problem 29: a job in the city, a five-year contract, a home less than 16 blocks away, and a desire to travel to work along a variety of routes.

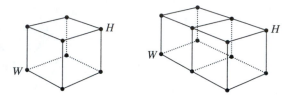

What is the nearest that a prairie dog could live to its place of work and have access to enough different routes?

How many years could a prairie dog work at this job if its home were located at a site that allows for a maximum number of routes?

Working on Mini-Projects in Teams.   The following are suggested guidelines for working on mini-projects in teams. These guidelines include details for presenting each team's results to the rest of the class, writing up the details of the project in a paper, and evaluating the experience.

Choosing the Problem   Look over the preceding collection of mini-projects. Try some of the projects that appeal to you. Look around for other persons in your class who are interested in working on a project that is attractive to you. You will want to be a member of a three- or four-person team. Everybody in the class will be on one team only. Each team will work on a different project. (That will maximize your exposure to different projects and approaches to solving problems.)

Solving the Problem   Since you will have the benefits of working with other members of the team, you can divide up the work. For example, if the problem calls for investigating several different cases, have one person handle *these* cases, another person *those* cases, a third person the *other cases*, and so on. Then the results of these investigations can be brought together to be considered by the group as a whole.

You could also deal with the problem by having every team member go off and try to solve the problem by him- or herself. Then come together to share notes on how things are progressing.

However you approach the problem, other members of the team can listen to your observations and hear your arguments. What you have to say may give them ideas for new lines of investigation. And you may get similar inspiration from what they have to say. They may also catch, before it's too late, some faulty reasoning. And they will commiserate with you if you are stymied.

Presentation   Count on having your team make a presentation to the other members of the class. All the presentations should take place over a one- to two-week period—a sort of festival of mini-projects. Each presentation should last about twenty minutes. The purpose of the presentation is to communicate something about the problem to the rest of the class: the nature of the problem, some of its features that might not be evident from a superficial look, some aspects of the group's investigations, and a solution that makes sense. You want to catch the interest of the persons in your audience. You want to convey to them the excitement of the hunt. Because of time limitations you may want to use the overhead projector with transparencies prepared ahead of time. You may want to use diagrams or tables on poster board prepared in advance. It would be useful to hand out a single-page summary of your investigation at the beginning of your presentation.

It is important that you go through the entire presentation from beginning to end (speaking out loud, standing up, using blackboard or overhead, . . .) before you present your project in class. You might want to do this more than once. You might want to try out your presentation with your teacher. Another group to try out your presentation with is a group of student-friends not in your class.

Paper   Each team will compile a paper detailing the results of the project. In addition to the items covered in the presentation, you can include details that you were unable to include in the presentation, due to time limitations. You can add more key examples. You can provide arguments missing in the presentation. You can show additional charts and diagrams. The paper is a place to provide biblio-graphic sources, if appropriate.

In any case, the purpose of your paper is to communicate the results of the team's investigations. You want it to be clear, complete, and convincing.

Participation   It is important that each team member participates in all decisions of the group and all aspects of the project. It is the responsibility of the group to make sure that each individual understands what is going on; it is an individual's responsibility to say when he or she doesn't understand an approach or argument proposed by other team member(s).

It may be useful to select a leader to keep the group organized and on task, to set up meetings, to divide up the work, ... The role of such a person is not to pre-empt the team by dominating discussions and making all the decisions.

Evaluation   There will be three types of evaluations. The first will be an evaluation by your peers of the presentations. Each of you, at the time of a presentation, will be asked to critique the presentation using the four C's as criteria: Clarity, Completeness, Correctness (is it Convincing?), and Creativity. (See more details on the four C's in Chapter 9, under *Project Evaluation*.)

The second evaluation will be personal. Some questions to answer in this evaluation follow. From your point of view, what role did you and the others in your team play in your project? Did the team work well? Was working on the team problem a valuable learning experience? How? Are you pleased with the outcomes of the project? What would you do differently the next time you are involved in a team project?

Your teacher will do the third evaluation using the four C's mentioned above.

✳ *Acknowledgment*

A discussion of the Five Planes Problem appears in *Induction and Analogy in Mathematics*, by Georg Pólya, pp. 43–52. Solving the problem is the plot of *Let Us Teach Guessing*, an hour-long film also starring Georg Pólya. The author acknowledges his deep indebtedness to these sources. A look at either of these would provide a great retrospective to our solving of the problem.

✳ **References**

Mason, John, *Thinking mathematically*. Reading, Mass.: Addison-Wesley, 1985.

O'Daffer, Phares, and Stanley R. Clemens, *Geometry: An investigative approach*. Menlo Park, CA: Addison-Wesley, 1997.

Pólya, Georg, *How to solve it: a new aspect of mathematical method*. Princeton, NJ: Princeton University Press, 1945.

Pólya, Georg, *Let us teach guessing*. Video tape. Washington, DC: Mathematical Association of America, 1970.

Pólya, Georg, *Induction and analogy in mathematics, Vol I of Mathematics and plausible reasoning*. Princeton, NJ: Princeton University Press, 1954.

Stevenson, Frederick, *Exploratory problems in mathematics*. Reston, VA: National Council of Teachers of Mathematics, 1993.

2 Episodes in the Measurement of Length, Area, and Volume

Three important types of geometric measurements are length, area, and volume. Three useful tools for making these measurements are *formulas, dissections,* and *approximations.* You are acquainted with formulas associated with a variety of nice shapes. You know that dissection is a strategy for dealing with a shape that isn't nice: dissect the shape into nice ones then use the formulas. What you may not know (or once knew and have forgotten) is how many formulas come about because of a dissection argument. You may have seen approximation as a mathematical tool for making these measurements—perhaps in a calculus course. Chances are you are not as comfortable using it as you are using formulas and dissection. You may not remember that some formulas come about because of an approximation argument.

What I want to do in this chapter is examine some familiar measurement formulas and see if we can make sense of them—at a fairly basic level, at a level deeper than sheer memorization, and at a level appropriate for middle- and high school understanding. Along the way I want us to begin to think of dissection and approximation as natural tools for solving measurement problems. I also want us to experience some situations in which a formula might not be the best tool for making a measurement.

To do all of this I want to look at several real-life measurement problems. I want us to use common sense in solving each one. This might mean shedding those mathematics classroom behaviors which tell us that there is a single, "approved" method of solving a particular problem.

This chapter is not a comprehensive treatment of length, area, and volume. Rather, it features certain high points in the measurement saga—finding the circumference of a circle, the area of a circle, the volume of a sphere, and the surface area of a sphere. These episodes are sufficiently rich in themselves to illustrate several approaches to measurement and how they intertwine. Moreover, they also are splendid occasions to illustrate the use of analogy as a strategy for solving problems.

MEASURING LENGTH

✵ *The Odometer Problem*

Your car's odometer calculates the number of revolutions your front wheel makes. From this it calculates how far the car has gone. To do this it needs to know how far the car goes when the wheel makes one revolution. How do you find out?

!

STOP! Try this before reading on.

One Solution to the Odometer Problem.　You could do it in the following way. Mark the road where the front tire is tangent to it. Mark the tire at this point of tangency. Drive the car slowly along the road in a straight line, stopping when the mark on the tire touches the road again. Mark the road at that point. Measure the distance between the two points marked on the road.

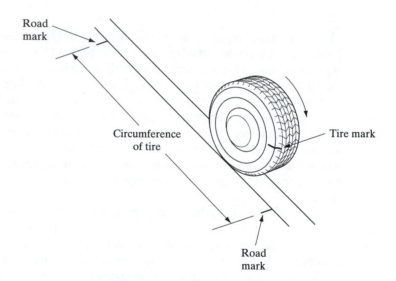

Road mark

Circumference of tire

Tire mark

Road mark

Other Solutions to the Odometer Problem.　Some people say "When the wheel makes one revolution as the car travels along the road, the car will have travelled the length of the circumference (or the perimeter) of the wheel." Do you believe this? If what these people say is true, then we've described one way to measure the circumference of the wheel (that is, drive the car, etc.). Are there other ways?

Another way is to wrap a string exactly once around the wheel's circumference. Then measure the length of the string with a measuring tape.

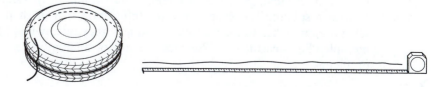

A third way—if you have no string but do have a reproduction on paper of the cross section of the wheel—is to approximate the circumference of the circle by straight line segments. Then measure the segments and add up the result.

A fourth way—and this is the route most of us well-behaved mathematics students would try first—is to use the formula for the circumference of a circle:

$$\text{(circumference of circle of radius } R) = 2\pi R$$

Problem for Right Now!

Before moving on, think about the four methods discussed in the solution to the odometer problem. What method would you use? If you were a person with less mathematical sophistication than you now have, what method would you use? Which method is more valid? What are advantages and disadvantages to each of the methods? What does the fourth method need that the others don't?

✳ Making Sense of the Formula for the Circumference of a Circle

It would be nice to make sense of the formula for the circumference of a circle. Before we do this, we ought to try answering the following question: Why is the formula a good thing? Think about this for a moment before reading on.

If you know the radius of a circle and you know π, then it's a matter of using your calculator to calculate the circumference. You don't need to get your hands dirty with string or other messy things. But when do you know the radius? If you've constructed the circle yourself, then measuring the radius is easy. To construct a circle, you take a piece of string. Tie a pencil in one end and a nail in the other. Pound the nail in and trace the circle with the pencil.

The circle is the set of all points whose distance from the nail is equal to the length of the string from nail (the circle's **center**) to pencil. This length is called the **radius** of the circle. Put another way, if you know the center of the

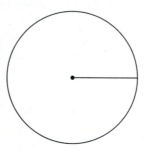

circle, then the radius is the distance from the center to a point on the circle (and it doesn't matter which point on the circle). We're more adept at this straight line measurement than we are with curvy ones.

Problem for Right Now!

What if you don't know the radius, or it's difficult to find? Take a full silo—a cylindrical grain storage bin; you want to find its circumference. Or take a conical-shaped pile of gravel; you want to find the circumference of the circular base of the pile.

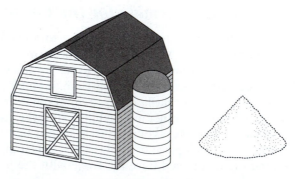

A formula for the perimeter P of a square is

$$P = 4S$$

where the length of one side of the square is S. Why is this formula useful?

The formula for the circumference of a circle can be useful. But it has its limitations. Actually, it's much more useful than we have admitted so far. To appreciate this, you need to know more. Wait and see!

At this point let's leave uses and limitations behind and try to understand what the formula is and why it works. To do this we need to forget we know the formula and try to figure out a way to calculate the circumference of a circle assuming we know its radius.

Is there anything obvious we can say about the relationship between the radius and the circumference? If the radius gets bigger, what happens to the circumference? It gets bigger, too. That seems reasonable. Let's see if we can be more precise and say just how much bigger. Take two circles—with different radii—and give them the same center. Approximate the circumference of the larger circle using straight line segments:

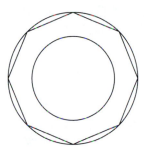

This was one of our methods, remember? To get the radius into the picture in a way that's connected with our approximation, draw the radii connecting the center to the endpoints of our approximating segments. Now connect the points of intersection of these radii with the smaller circle:

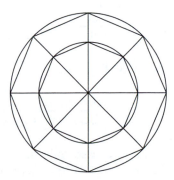

This gives us a bunch of lengths whose sum approximates the circumference of the smaller circle.

Next, look at a pair of adjacent radii in this picture, corresponding to two end points of a segment:

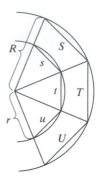

We get a pair of triangles! Triangles are nice (we know them). Is there anything we can say about these two triangles? Yes! They're similar! The ratio of corresponding sides (smaller to larger) is equal to r/R. This means that the ratio of the length s (approximating the circumference of the smaller circle) to the length S (approximating the circumference of the larger circle) is also equal to r/R. In other words,

$$\frac{s}{S} = \frac{r}{R}$$

This must be true of any other pair of corresponding segments approximating the circumferences of the two circles:

$$\frac{u}{U} = \frac{t}{T} = \frac{r}{R}$$

Thus it must also be true of the approximations of the two circumferences:

$$\frac{\text{approximation of circumference of circle of radius } r}{\text{approximation of circumference of circle of radius } R} = \frac{r}{R}$$

It seems reasonable, then, that

$$\frac{\text{circumference of circle of radius } r}{\text{circumference of circle of radius } R} = \frac{r}{R}$$

Problem for Right Now!

Why does it seem reasonable that this last equality follows from the previous equality involving the ratio of approximations?

The last equation can be rewritten as

$$\frac{\text{circumference of circle of radius } r}{r} = \frac{\text{circumference of circle of radius } R}{R}$$

$$= K$$

Thus, the circumference of a circle of radius r is equal to r times a fixed number K—the fixed number is the same for circles of all sizes. Fine. What is the fixed number? In the equation, let $R = 1$ unit. Then

$$\frac{\text{circumference of circle of radius } r}{\text{circumference of circle of radius } 1} = \frac{r}{1}$$

or

circumference of circle of radius $r = r \cdot$ (circumference of circle of radius 1)

So the fixed number is the circumference of the circle of radius 1. What is that? (Secretly, we know what it should be.) It's something we can measure, or approximate, by any of the techniques we have described. Haven't we come full circle? (!) Didn't we want to avoid making crude measurements? Yes. But we have reduced the problem of finding the circumference of all circles to making a one-time-only measurement: measure the circumference of a circle of radius 1. Do it very carefully! Then you can use the results of that single measurement to find a formula that gives the circumference of any other circle in terms of its radius. If $C(r)$ denotes the circumference of a circle of radius r, then the number we have measured is $C(1)$ and we have the following formula:

$$C(r) = C(1)r$$

Traditionally, we call 2π the common ratio $C(r)/r$ and write $C(r) = 2\pi r$.

✳ *A More Convincing Argument*

In the discussion above, we have given an argument that makes the formula $C(r) = C(1)r$ plausible. In this section we will outline how one might add details to that argument in order to make the outcome more certain; i.e., we seek a more convincing proof of the formula.

The first thing we have to do is to come up with a good mathematical definition. A real-life, operating definition of the circumference of a circle is to take a piece of string, fit it tightly around the circle, straighten it out, and measure it. We can carry this out in the real world for any circle we can stretch a string around. But it doesn't give us a formula which, in our case, relates **two** measurements—radius and circumference—in a precise way and for all possible radii: no matter what the radius r is (precisely), there is a single number $C(1)$ that works for all r such that $C(r)$ is precisely $C(1)r$. To relate these two measurements with this kind of precision, we need to describe (mathematically) how we are to measure the length of a curve, given that we can measure (mathematically) lengths of straight line segments. We know what the expressions "fit it tightly" and "straighten it out" mean in real life but, at this point, it would be difficult to use them in coming up with a mathematical definition of the length of a curve in terms of things we know: lengths of straight lines. However, another real-life method for measuring the circumference of a circle might help us. Take a small ruler of length L. Start with a point P on the circle. Find the point P_1 on the circle of distance L from P, then the point P_2 on the circle of distance L from P_1, and so on around the circle until you reach point P_n on the circle (to the "left" of P) whose distance is less than or equal to L.

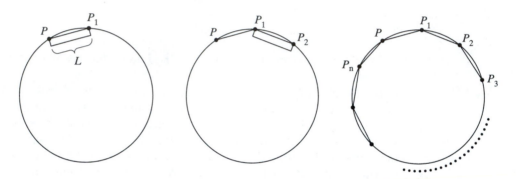

You feel that the circumference of the circle is bigger than or equal to nL. You feel that your estimate gets closer and closer to the "true" value as L becomes smaller and smaller. In formal language, we **define** the circumference of the circle to be the limit of these estimates as L goes to 0.

We can systematize this method further and also brings it into line with the discussion preceding this section by selecting only those L whose lengths are the sides of regular, inscribed polygons. In fact, let $P_n(r)$ denote the perimeter of a regular n-gon inscribed in a circle of radius r. As before, let $C(r)$ denote the circumference of a circle of radius r. Then, by our definition, we have that $\lim_{n\to\infty} P_n(r)$ $= C(r)$. Also, by the argument preceding this section, we have that $P_n(r) = P_n(1)r$. Thus,

$$C(r) = \lim_{n\to\infty} P_n(r) = \lim_{n\to\infty} P_n(1)r = r\lim_{n\to\infty} P_n(1) = rC(1)$$

This doesn't tell us what $C(1)$ is, however. Problem 1 gives a way, using these ideas, to come up with good approximations for the number $C(1)$.

Problem 1

Archimedes approximated the value of $C(1) = 2\pi$ using the following method. Take a circle of radius 1. In it inscribe a regular hexagon of perimeter h; around it circumscribe a regular hexagon of perimeter H. Then $H > C(1) > h$. (Why is this so?)

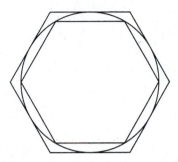

Next, by doubling the sides of the regular hexagons, Archimedes creates regular 12-gons, one with perimeter t inscribed in the circle, the other with perimeter T circumscribing the circle. Then $H > T > C(1) > t > h$. (Why is this so?) He keeps doubling the sides of the inscribed and circumscribed polygons and finally arrives at inscribed and circumscribed regular 96-gons of perimeters n and N, respectively. He gets $N > C(1) > n$ and from that concludes $22/7 > \pi > 223/71$.

Carry out the first two steps of Archimedes' procedure, i.e., find h, H, t, and T.

Problem 2

A certain circular field is irrigated by a pipe that rotates about its center. You want to build a fence around it. How much fencing do you need? How do you find out?

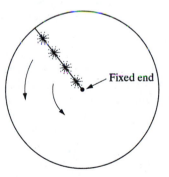

Fixed end

Problem 3

(a) You own a circular flower bed that you plan to enlarge by 1 meter (m) all the way around. The present flower bed has fencing that you plan to remove and use for the expanded flower bed. You will need additional fencing. How much? Comments?

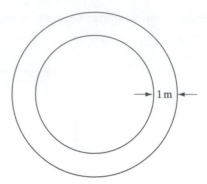

(b) You order a piece of string 24,900 miles (mi) long, which you assume is just long enough to encircle the globe exactly, at the equator. You take this string and fit it snugly around the earth, over oceans, deserts, and jungles. Unfortunately, you find that there has been a slight error: the string is just a yard too long! To overcome the mistake, you decide to distribute the extra 36 in. evenly over the entire 24,900 mi by propping it up equally at all points on the equator. Naturally, it will never be noticed, but you are interested in how far the string will stand off from the ground at each point. How much do you figure? Comments?

Problem 4   **(a)** An automobile has wheels that are 18 in. in diameter. How many revolutions will each wheel make on a 1-mi trip?

(b) An automobile's speedometer measures the rate at which the drive shaft of the car is rotating, and, through the differential, the rate at which the wheels are turning. The speedometer is calibrated to be accurate when the tires are new (9/32-in. tread depth) and the air pressure is 30 lb/in^2. Now, the 18-in. diameter tires have worn to 1/16-in. tread depth. With the air pressure still at 30 lb/in^2, what is the error in a speedometer reading of 30 mph?

Problem 5   A trundle wheel is a wheel attached to a handle that guides the wheel as it rolls along the ground. The wheel has a circumference equal to 1 m, with decimeters (tenths of a meter) marked along the circumference of the wheel. It is used as a device for measuring lengths. Roofers use it. Street pavers use it. To use it, you place the "zero point" of the wheel at the beginning of the path you want to measure. Then you roll the wheel along the path. A clicking mechanism has been built into the wheel so that each time you cover a meter in distance—that is, each time the wheel makes a full revolution—the wheel makes a clicking noise. You want to make this trundle wheel. As you move the wheel along the path, you count the number of clicks. The length of the path is the number of clicks plus the number of decimeters indicated by the position on the wheel that rests on the end point of the path. You want to construct such a wheel. What will its diameter be? What will be the angle, measured from the center of the wheel, between consecutive decimeters on the circumference of the wheel?

An eighteenth-century
trundle wheel (perambulator)

Problem 6   A certain curved section of a highway is an arc of a circle—a very large circle. Highway safety folk know that a safe speed limit for driving along this curved section depends on the radius of the circle. The smaller the radius, the lower the safe speed limit. How would you find the radius of the circle corresponding to this curved section?

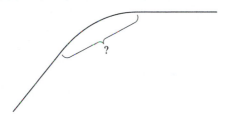

Problem 7   You design and make lamp shades. One such lamp shade is shown.

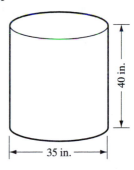

(a) Sketch a flattened pattern for this lamp shade. On the pattern mark in the dimensions needed for assembling it. The pattern should allow for a 1-in. overlap when gluing the pattern together.

(b) You plan to add trim to the top and bottom rims of the lamp shade. What length of trim will you need?

Problem 8   (With thanks to Marilyn vos Savant . . .) There are two circles. The larger circle has circumference exactly four times the circumference of the smaller circle. Roll the smaller around the edge of the larger circle. By the time the smaller circle returns to its starting position, how many complete rotations will it have made? Comments? Generalizations?

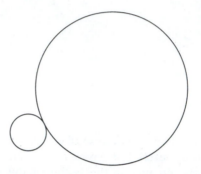

Problem 9   A segment of the circumference of a certain circle is shown below. It is called an arc of the circle. You are interested in the length of this arc. Find it.

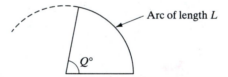

Arc of length L

$Q°$

A circumference of interest to all of us is the circumference of the earth, that is, the circumference of a great circle of the spherical earth. The equator is one such circle. A north–south meridian (through North and South Poles) is one.

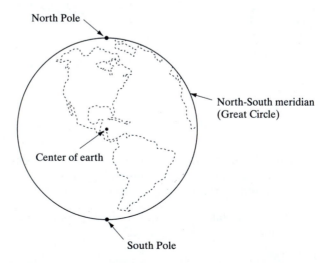

North Pole

North-South meridian
(Great Circle)

Center of earth

South Pole

How do you find the circumference of one of these circles? (String is out of the question. Rolling the circle along a flat piece of paper—are you kidding? Approximating it with straight line segments? Ridiculous. Use the formula? We'd need to know the radius. How do we find that? Phooey. Some kind of cleverness is needed.)

The Alexandrian mathematician, Eratosthenes (c. 276–c. 194 B.C.), provided this cleverness. The Greeks had already hypothesized that the earth was spherical. Eratosthenes's scheme depends on the following relationship between arc length and the measure of a certain angle, a relationship you may have deduced in working out the previous problem.

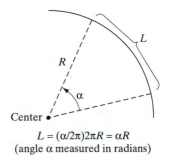

$$L = (\alpha/2\pi)2\pi R = \alpha R$$
(angle α measured in radians)

Eratosthenes discovered that the Egyptian cities of Alexandria and Syene were on a single north-south meridian, a great circle passing through the north and south poles. He was also able to measure the distance between the two cities along this meridian. Thus he knew L in the preceding formula. His idea was to find α, solve for R in this equation, and, from there, find the circumference using the familiar formula. But how was he to measure α?

Eratosthenes also knew that at noon on the summer solstice, the sun's rays would fall to the bottom of a certain well in Syene. At that moment, the sun's rays and the great circle from Alexandria to Syene would look like this:

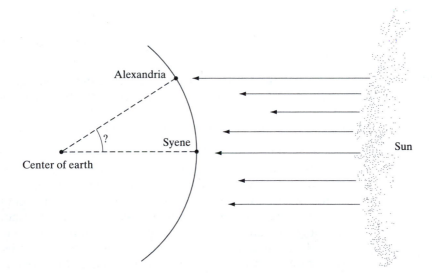

Eratosthenes argued this way: "Since the sun is so far away from the earth and is so big, its rays are essentially parallel. That means that the angle I want to measure is equal to the one marked in the following diagram:

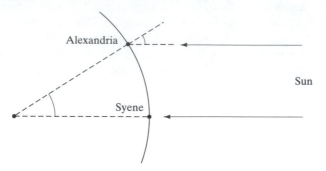

(Why is this?)

"To measure that angle, at noon on June 21 I will place a stick perpendicular to the ground at Alexandria. Then the rays of the sun will make an angle with the stick, and that's the angle I want to measure. To measure that, I measure the shadow the stick makes on the ground . . ."

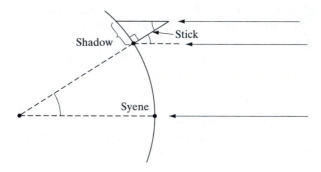

Problem 10
Eratosthenes still hasn't measured the important angle. What else does he need to do?

Problem 11
Eratosthenes knows that noon (the instant when the sun is highest in the sky) occurs at Syene when the sun's light reaches the bottom of the well. It's important that the measurement at Alexandria be made at the same instant, i.e., at noon. (Why?) How does Eratosthenes know when noon occurs at Alexandria?

Problem 12
My son Adam validated Eratosthenes's measurement of the circumference of the earth with a little help from his friends. A well, such as the one at Syene, into which the sun directly shines at summer solstice was not necessary. He found two locations on the same north-south meridian in the United States and measured the distance between the two locations using the scale of a map. One morning, he and a friend drove to one location; two other friends drove to the other location. Exactly at solar noon (when the sun is at its highest point in the sky), each team, using a stick perpendicular to the ground, measured the angle

of the rays of the sun at that location. A diagram showing the details of this experiment is shown. Adam claimed that angle *A* minus angle *B* is equal to angle *C*. Why is this? How did he use this information to calculate the circumference of the earth?

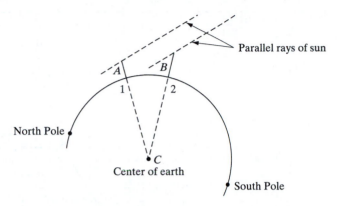

(Eratosthenes's measurement was made on summer solstice. Should Adam's measurement be made on any special days?)

Problem 13

Construct the clinometer shown below. A clinometer is a simplified version of both a quadrant, a medieval instrument for measuring distances, and a sextant, an instrument for locating the position of a ship. The device is used to measure

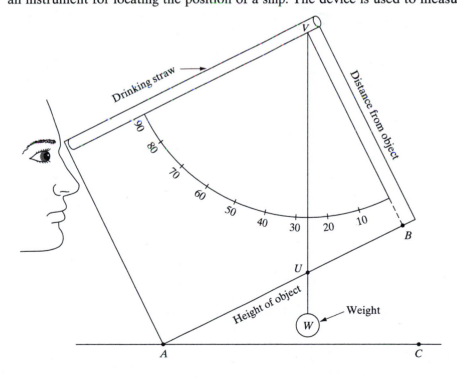

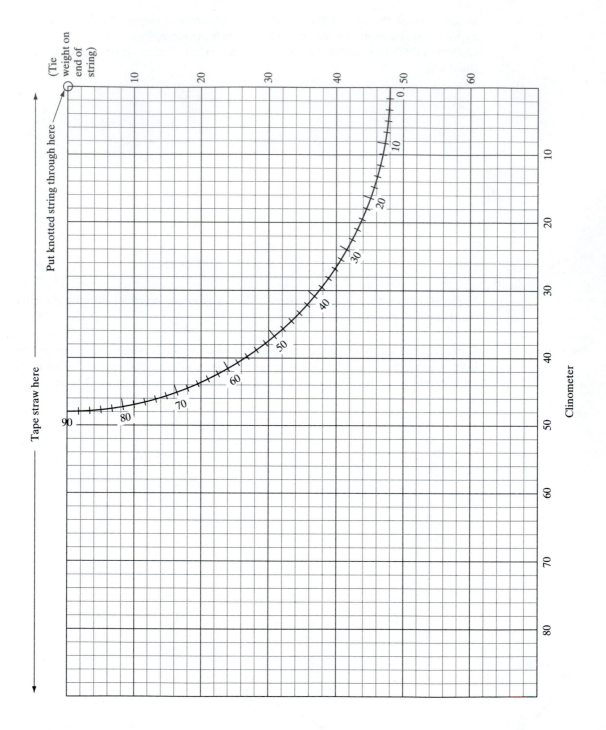

Clinometer

56

heights of tall objects (buildings, mountains, trees, etc.) as well as angles of inclination. On the clinometer is printed a circular arc, graduated in degrees from 0° to 90°. To use it, you sight a feature through the straw, then read the measure of angle *BVU* on the graduated arc.

Angle *BAC* is the angle of elevation of the clinometer. Angles *BAC* and *BVU* are congruent. (Why is this?) This fact is what makes the clinometer so useful. Here is an example.

You want to find the height of the tree shown in the following picture. You use a clinometer to find the angle of elevation from your eye level to the top of the tree. You also measure the distance along the ground from where you make the observation to the base of the tree. Find the height of the tree in two ways. Use trigonometry to find the distance *RT*. Or use the fact that triangle *BVU* is similar to triangle *TER*. (Why is this last statement true?)

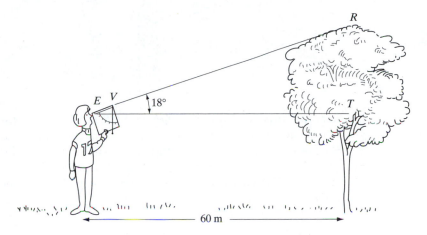

60 m

Problem 14

Your latitude in the northern hemisphere is determined in the following manner. The great circle passing between you at *P* and the North Pole cuts the equator at a point *Q*. Your latitude is the measure of the angle *POQ*, where *O* is the center of the earth.

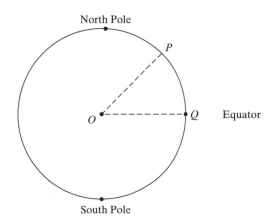

How do you determine the measure of this angle? One way is to use the clinometer—see previous problem—and find the angle of elevation of the North Star above the horizon. The angle of elevation is your latitude! Why is this? The following diagram may help to explain what is going on.

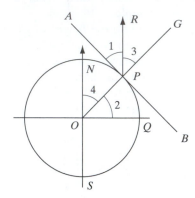

You are at point P and rays ON and PR point to the North Star. Because the North Star is so far away, ON and PR are parallel for all practical purposes. Furthermore, line OG is perpendicular to line AB, your sight line to the horizon. Your latitude is the measure of angle 2; the angle of elevation that you measure is angle 1. These two angles are the same! Why? Use this method to find your latitude.

MEASURING AREA

Our main objective in this section is to discuss methods for finding the areas of circles. To provide a context for doing this, let's recall some typical early events in learning about area. We begin with a problem.

✳ *The Garden Plot Problem*

Several families have plots of land on which they grow potatoes. In the accompanying scale drawings of these plots, each little square is a meter on a side. Each family can expect a yield of $7 from each little square. What is the yield for each plot?

!

STOP! Try this yourself before reading on.

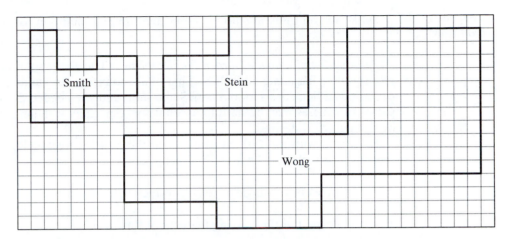

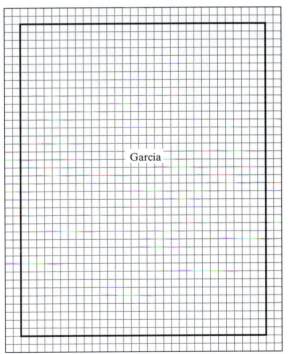

One Solution to the Garden Plot Problem. You know that to calculate the yield you must know the number of little squares in each plot. You know that the latter number is the area of the plot. (The square—a meter on a side—is the unit of area, a square meter.) For example, take the Smith family plot. You count the number of squares and get 33. The area of the Smith family plot is 33 square meters (m²). At $7 per square meter ($7/m²), the yield is 7 × 33 = $231.

You can do the same thing with the other plots. Count the squares.

Problem for Right Now!   What do you think about the "count the squares" solution for the other plots?

Another Solution to the Garden Plot Problem.   The above solution to the garden plot problem has some difficulties. First of all, counting the number of squares can be tedious if there are a lot of them. Some short cuts are needed. Second, fractional pieces of squares may be hard to account for, especially if there are a lot of them or if accuracy is desired.

Let's deal with the first difficulty when the plot is a rectangle. You know that there is a formula for the area of a rectangle:

$$\text{area of rectangle} = \text{length of rectangle} \times \text{width of rectangle}$$

You appreciate the time-saving features of this formula.

Problem 15   The formula makes sense when the lengths of the sides of the rectangle are whole numbers. Why is this? The formula also makes sense when the lengths of the sides are fractions. Why is this?

Let's deal with the second difficulty (fractional pieces of the unit square) when the plot is a right triangle. To find the area of a right triangle you put it inside a rectangle:

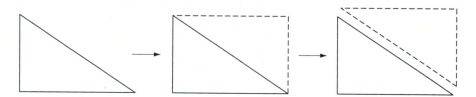

You notice that the rectangle's area (which you know) is the area of the triangle plus "the rest." You notice that the rest is another triangle congruent to the one you are interested in. Areas of congruent shapes must be the same. So the area of the rectangle is twice the area of the original triangle. Equivalently, the triangle is half the area of the rectangle. This gives you a formula for the area of a triangle.

✳ *Area Principles and Strategies*

Finding the area of a right triangle gives you some insights which might be useful in other situations.

Area Principles.

• Dissect a planar shape into pieces. The area of the shape is the sum of the areas of the pieces.

• Congruent shapes have the same area.

These principles provide you with strategies for finding the area of a shape:

• Dissect the shape into pieces whose areas you know.

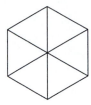

• Dissect the shape into pieces which you then arrange into a shape whose area you know.

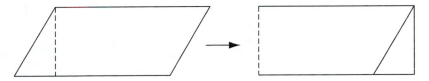

• To the shape of interest, add additional shapes (whose areas you know) to form a larger shape whose area you know.

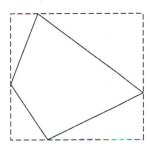

• Make several copies of the shape. Arrange the copies to make a shape whose area you know.

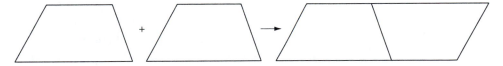

Problem 16   Use the principles, strategies, and formulas for rectangles and right triangles to derive the traditional formulas for

(a)   the area of a parallelogram

(b)   the area of any triangle (not just a right triangle)

(c)   the area of a trapezoid

Problem 17

To the Greeks, the Pythagorean theorem was really a theorem about area: Given a right triangle, the area of the square on its hypothenuse is equal to the sum of the areas of the squares on its other sides.

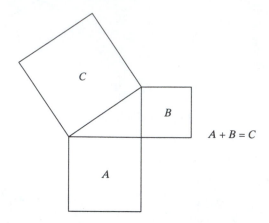

$A + B = C$

Take a right triangle whose sides other than the hypothenuse have lengths a and b. The two dissections shown below—both of a square whose side has length $a + b$—give you a proof of the Pythagorean theorem. Why is this?

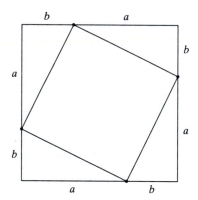 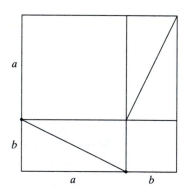

Problem 18

You have been invited to Triangle Land to help design school curriculum materials in mathematics. You have heard that the country's approach to measuring is different than your own. You feel that it wouldn't be a good idea to dump your own notions of area on them, but that you should take what they have and build from there. You decide to learn as much about their system as possible. First you discover that their system for measuring lengths is just like yours: they use a meter as the unit of length! However, in their system for measuring area there is a twist. The traditional unit of area is an equilateral triangle each of whose sides is a meter in length. This unit is called the "triangular meter," (tm).

In Triangle Land, it seems that a triangular grid is preferred to our rectangular one and plots of land tend to favor angles of 60 and 120 degrees. Some garden plots in this country are shown in the following diagram.

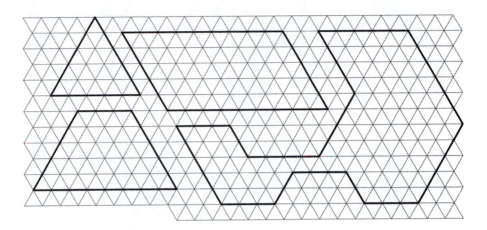

To get a feel for calculating area in Triangle Land, you decide to find the areas of each of the plots in number of triangular meters (tm's). What do you get?

Based on this experience, as well as on your experience in your own World of Squares, you think: "In my world, there are shapes whose areas are easy to figure out. For example, a rectangle whose sides have whole number lengths is easy: count the length of two sides and multiply them together. What would be an analogous shape in Triangle Land?" You decide to try an equilateral triangle, the length of whose side is n meters, where n is a whole number. What is its area in tm's?

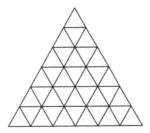

Energized by your success, you think: "What are some other shapes whose Triangle Areas might be easy to calculate?" You decide to try a parallelogram having one of its angles equal to 60° and whose sides have whole number lengths. What is its area in number of tm's?

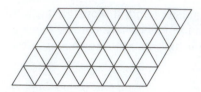

Once again, you are elated by figuring out some of the mysteries of this system. In travelling around Triangle Land, you notice that a lot of houses are in the shape of a regular hexagon and figure that the regular hexagon must be a shape whose area is easy to determine. You decide to find out for a regular hexagon *n* meters on a side. What is its area in tm's?

You realize through all of this that you have been using the area principles you used in the World of Squares (those on pp. 60–61). One thing you haven't worried about is whether your formulas work when a length measurement is a fraction. You decide to pause and look at an equilateral triangle the length of whose side is the fraction *a/b*. You want to know whether or not the formula you got earlier (when the length of a side is a whole number) works in this case and, if it does, what's a good argument for it. What do you find out? And, if the formula works, what's your argument?

You decide to test what you have found with some of the locals. "Excellent," they say, "you seem to have caught on to the spirit of our system. Here is something that might interest you. We have heard that in your world you rely heavily on what you call a 'right' angle. Here, although 90° angles are important, we tend to use the 60° angle more. For example, with a rectangle—a four-sided figure with all its angles equal to 90°—we make two length measurements for calculating its area: its base (as you do) and its 60° 'slant' height—along a line that makes a 60° angle with its base. For your next project, why not use these two measurements to find the area of a rectangle?" You decide to take them up on their suggestion. What do you conclude? Why?

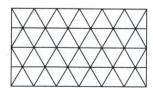

Finally, you decide to seek formulas (and arguments justifying them) for the areas of any old triangle, any old parallelogram, and any old trapezoid using what you know and measurements that would be "natural" to Triangle Landers. What formulas do you come up with? And what are your arguments for them?

✳ *The Circular Garden Plot Problem*

Find the area of the garden plot shown in the following scale drawing. It's a plot irrigated by a rotating pipe. Each square in the grid is a square meter. You don't know anything about a formula for the area of a circle. How would you find the area?

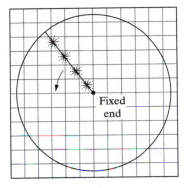

STOP! Try this before reading on!

A Solution to the Circular Garden Plot Problem. Having the grid right before you suggests that you count the number of whole squares inside the circle. After that, you estimate what fraction each piece of square is. You add the estimations to the whole number to get an estimated area. What is your estimate?

Another Solution to the Circular Garden Plot Problem. You are not happy with your estimate. You think, "Maybe there is a better way." You know that this is not like a shape you have dealt with before, a shape having straight line segments for its perimeter. The circle is different; it has a curvy perimeter. But the circle was different before, when you had to find the *length* of its curvy perimeter. You solved that problem. Maybe its solution might help here. There, you drew the following picture.

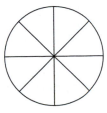

The picture shows a dissection of the circle into pie-shaped wedges. The area of the circle is the sum of the areas of the wedges.

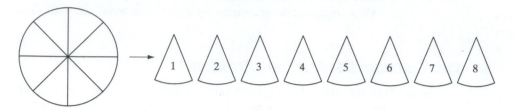

Moreover, each wedge is approximately a triangle—and you know the area of a triangle. So the area of the circle is approximately the sum of the areas of the triangles.

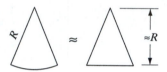

Now the height of each triangle is approximately the radius R of the circle. So the area of the triangle is approximately

$$\tfrac{1}{2}(\text{base of triangle})R$$

Thus,

$$\text{area of circle} \approx \tfrac{1}{2}(\text{base of triangle 1})\,R$$
$$+ \tfrac{1}{2}(\text{base of triangle 2})\,R$$
$$+ \cdot\,\cdot\,\cdot$$
$$+ \tfrac{1}{2}(\text{base of triangle } k)\,R$$
$$= \tfrac{1}{2}R\,(\text{base of triangle 1}$$
$$+ \text{base of triangle 2}$$
$$+ \cdot\,\cdot\,\cdot$$
$$+ \text{base of triangle } k)$$

But, the sum of the bases of the triangles is approximately the circumference of the circle. Thus

$$\text{area of circle} \approx \tfrac{1}{2}R \cdot (\text{circumference of circle})$$

The approximation to the area of the circle as a sum of areas of triangles gets better as the wedges become smaller, at which time the approximation to the circumference as a sum of the bases of the triangles also gets better. Thus the following exact formula is plausible:

$$\text{area of circle} = \tfrac{1}{2} R \cdot (\text{circumference of circle of radius } R)$$

Then, using the formula we got earlier for the circumference of a circle, we have the following.

$$\text{area of circle of radius } R = \tfrac{1}{2} R \cdot R \cdot (\text{circumference of a circle of radius } 1)$$
$$= \tfrac{1}{2}(\text{circumference of circle of radius } 1)R^2$$

As we noted earlier, one half of the circumference of a circle of radius 1 is traditionally denoted by the Greek letter π so that

$$(\text{area of circle of radius } R) = \pi R^2$$

Neat!

Problem 19 The discussion above provides an informal argument for the plausibility of the formula

$$(\text{area of circle of radius } R) = (\tfrac{1}{2})(R)(\text{circumference of circle of radius } R).$$

Add more details to make the argument more rigorous and thus provide a valid proof of the formula. A model for how to carry this out is given in the section entitled "A More Convincing Argument."

Problem 20 We discussed several methods for estimating the circumference of a circle: rolling the circle on a piece of paper . . . , putting a string around the circle's circumference . . . , approximating the circumference by a bunch of straight line segments . . . . What are some analogous methods for estimating the area of a circle? (Remember, the methods don't have to be sophisticated.) What do you think of these methods?

Problem 21 **(a)** Recall the cylindrical lamp shade from problem 6. Find out how much material is used to make the cylindrical lamp shade.

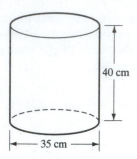

(b) You irrigate your alfalfa fields with pipes that rotate about a center. Thus land where alfalfa is grown consists of a bunch of circles. Some of the land will not be used. How should the circles be placed to maximize the use of the land for growing alfalfa? Two possible configurations for the circles are shown below.

(c) You run a fish canning operation. You want to know how much metal you'll need for 100,000 tuna cans, 1.2 in. high, 3 in. diameter. How to find out? (There will be wasted metal for the tops. Why? How much?)

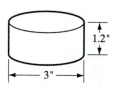

Problem 22   **(a)** Find the area of the sector of the following circle.

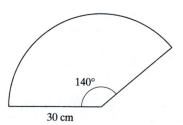

(b) Another lamp shade that you make is shown next in the form of a cone. Sketch a flattened pattern for it and label it with all dimensions necessary for assembling the real thing. Find out how much material is used to make the lamp shade.

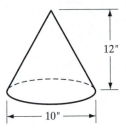

(c) Use the solution to part (b) to find a formula for the surface area of a cone.

(d) Yet another lamp shade you make is shown below.

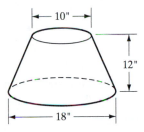

Sketch a flattened pattern for this. On the pattern mark in the dimensions needed for assembling it. The pattern should allow for a 1-in. overlap when gluing it together. How much material will you need to make the lamp shade?

VOLUME MEASUREMENT

Early experience with volume is similar to that for area. You start with rectangular boxes and soon acquire a formula for the volume of a rectangular box. If you replace each occurrence of the word "area" by "volume," each principle for area becomes one for volume. Similarly, each dissection strategy for finding areas becomes one for finding volumes.

After finding volumes of rectangular boxes, it's likely that you used the dissection strategies to find the volumes of right prisms, several examples of which are shown as follows.

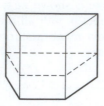 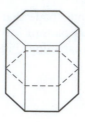 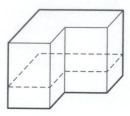

Each of these shapes has a base which rests securely on the ground. Each cross section of the shape formed by a horizontal plane parallel to the ground is congruent to the base. All parts of the shape lie directly above the base. The shape doesn't lean as the following one does.

(The latter shape is not a right prism. It is sometimes called an oblique prism.)
For example, let's find the volume of a right prism whose base is a right triangle.

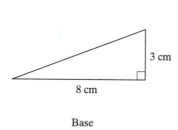

Base

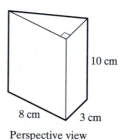

Perspective view

The similar, simpler problem of finding the area of a right triangle suggests an approach to finding the volume of this shape. You make two copies of the shape which you then fit together to make a rectangular box. The volume of the shape is thus one-half the volume of the box. Looking at the problem from another point of view, you notice that the volume of the rectangular box is the area of its base times its height. Thus, the volume of the triangular prism is the area of **its** base times its height.

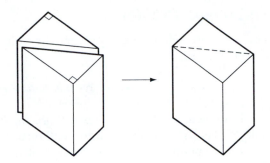

Problem 23   Using what you have done, show how to find the volumes of the following right prisms. Explain your methods.

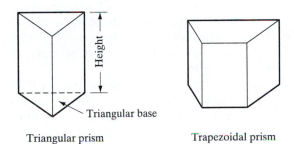

Triangular base

Triangular prism   Trapezoidal prism

Problem 24   Again, using what you have done, give an argument showing that the volume of any right prism having a base that is a polygon is the area of the polygon times the height of the prism.

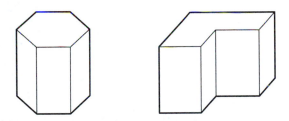

Problem 25   Justify (carefully) the usual formula for the volume of a cylinder.

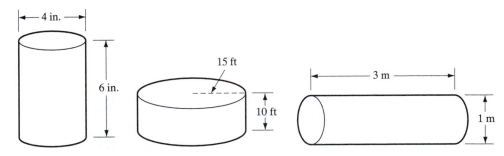

✳ *The Sphere Volume Problem*

Take a sphere. How would you find its volume? The method doesn't have to be sophisticated.

!   ***STOP! Try this before reading on!***

A Solution to the Sphere Volume Problem, Begun. When we see "sphere" we think "circle." The sphere seen from all directions looks like a circle. The circle is the set of all points in the plane equidistant from its center, while the sphere is the set of all points in space equidistant from its center.

As the circle was to planar shapes previously considered, so the sphere is to other three-dimensional shapes we've looked at. The circumference of the circle is "curvy"; it's not made of straight line segments. The surface of the sphere is "curvy"; it's not made of flat pieces. A shape we've considered that seems close to the sphere is the cylinder. But, unlike the skin of a sphere, the skin of a cylinder can be flattened out! (See problem 6.) Furthermore, the cylinder turns out to be a right prism and easy to deal with.

The circle appears to be a shape simpler than, yet analogous to, a sphere. Let's exploit the analogy in order to find the volume of the sphere. Area in two dimensions is analogous to volume in three dimensions. Perhaps the method of finding the area of a circle will give us ideas for finding the volumes of a sphere. To find the area of a circle we cut the circle up into wedges. The wedges all met at the center of the circle. Each wedge was approximately a triangle—something whose area we knew. The area of the circle was the sum of the areas of these wedges.

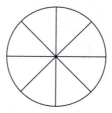

This suggests cutting the sphere into three-dimensional wedges. Each wedge might look like one of the following shapes.

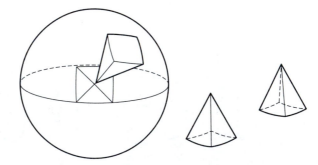

Cut the whole sphere into wedges like this. All the wedges would meet at the center of the sphere. The volume of the sphere would be the sum of the volumes of the wedges.

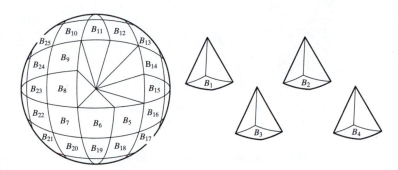

This reduces the problem of finding the volume of a sphere to finding the volume of a wedge. What is one of these wedges? If the wedge is small enough, it's roughly a pyramid.

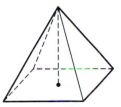

A pyramid is not a shape whose volume we know how to find, yet. Rats. The volume of a sphere is turning out to be more difficult than the area of a circle. With the circle, we approximated each wedge by a triangle, a shape whose area we knew.

Let's not give up. If we can figure out how to find the volume of a pyramid, then we can approximate the volume of a sphere with volumes of pyramids as we approximated the area of a circle with areas of triangles. It seems worth a try since finding the volume of a pyramid certainly looks easier than finding the volume of a sphere. After all, a pyramid has flat sides.

Problem 26   We discussed several methods for estimating the circumference of a circle: rolling the circle on a piece of paper . . . , putting a string around the circle's circumference . . . , approximating the circumference by a bunch of straight line segments . . . . What are some analogous methods for estimating the volume of a sphere? (Remember, the methods don't have to be sophisticated.) What do you think of these methods?

✳ The Triangular Pyramid Problem

Perhaps the simplest pyramid is a triangular pyramid. Its base is a triangle. An Egyptian pyramid (square-based) can be dissected into two triangular pyramids.

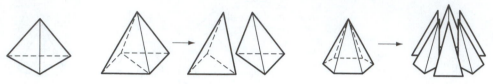

Triangular
pyramid

Square-based pyramid, dissected

Hexagon-based pyramid, dissected

You have a triangular pyramid. How would you find its volume?

!

STOP! Don't read on before trying this!

A Solution to the Triangular Pyramid Problem. You think, "The volume of a triangular pyramid in three-dimensions is analogous to the area of a triangle in two dimensions. How did I find a formula for the area of a triangle?

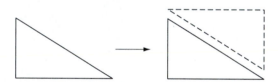

I used a dissection strategy for the triangle. Perhaps I can do the same thing with the triangular pyramid: make it part of a larger shape with a volume I know how to find. A triangular pyramid would fit very neatly into a triangular prism.

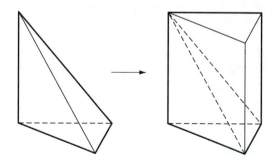

I know a formula for the volume of a triangular prism. I also know that the volume of the triangular prism is the sum of the volume of a triangular pyramid plus the volume of the "rest" of the prism.

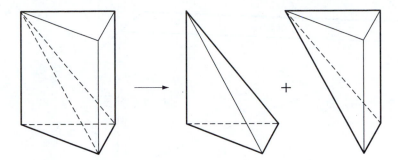

Hmmmm. The rest is a rectangular-based pyramid which I can dissect into two triangular-based pyramids.

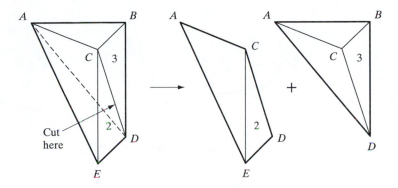

The volume of the triangular prism is the sum of the volumes of the three triangular pyramids."

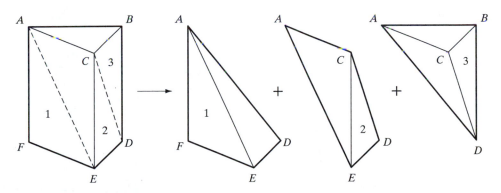

Problem for Right Now!

To convince yourself of what has happened so far and to follow the remainder of the solution to the triangular pyramid problem, cut out the following four patterns and assemble them into the three triangular pyramids and the triangular prism.

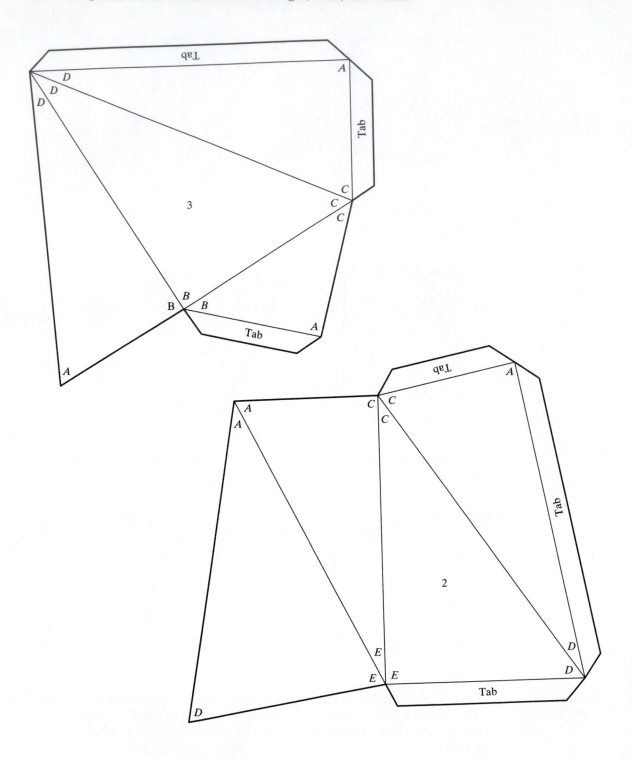

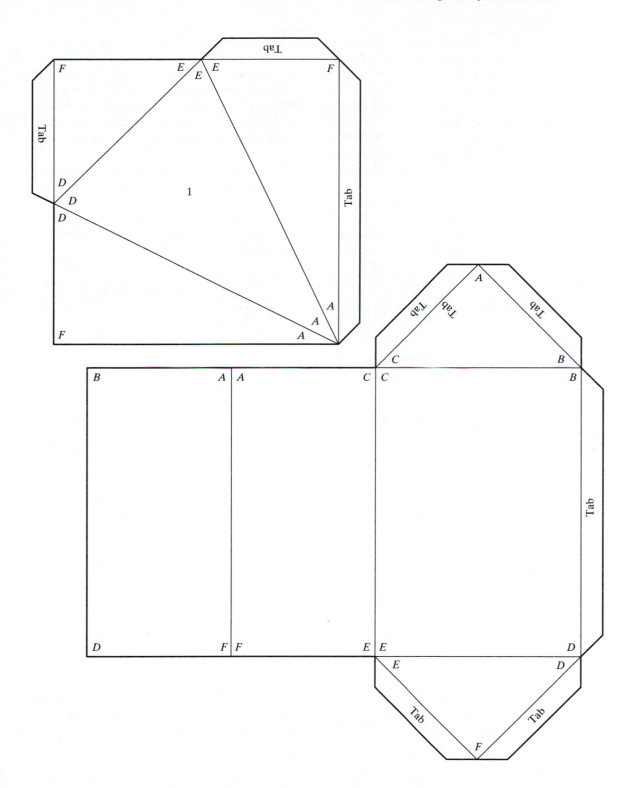

Continuation of a Solution to the Triangular Pyramid Problem. You continue: "So far, how does what we've done compare with finding the area of a right triangle? With the triangle, it and a copy of it fit together to make a rectangle. The areas of the two triangles are equal. Thus the area of the original triangle is half the area of the rectangle. With the triangular pyramid, it and two other triangular pyramids fit together to make a triangular prism. If the three triangular pyramids had equal volumes, then the volume of the triangular pyramid would be a third of the volume of the triangular prism. That would be very nice!

"There's a glitch. Whereas the two triangles making up the rectangle are congruent (and therefore have equal areas), the three pyramids making up the prism are not congruent. Foiled again.

"But let's look more closely at the three pyramids. Pyramids 1 (the original) and 3 have the same base and height. If I use different faces as bases, pyramids 2 and 3 have the same base and the same height above that base.

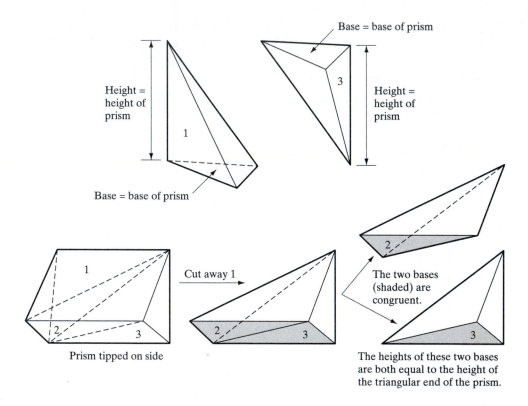

"In the analogous situation, triangles with the same base and height have the same areas. If two triangular pyramids with the same base and height had the same volume, the volume of the three pyramids would be the same and my problem would be solved."

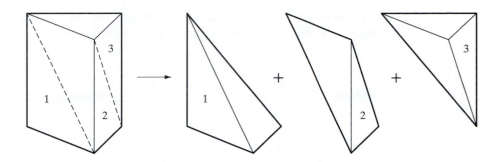

Problem 27   Relate the problem of finding the volume of a cone to finding the volume of a triangular pyramid.

A Solution to the Triangular Pyramid Problem by Layering.   To complete the solution to the triangular pyramid problem, we want to show that two triangular pyramids having congruent bases and equal heights also have equal volumes. The analogous problem in the plane is to show that two triangles having congruent bases and equal heights have equal areas. We know that we can show the latter using a dissection argument. That's how we got the formula for the area of a triangle. It turns out that a dissection argument can't be used to show the analogous fact for three-dimensional shapes. (This is not obvious. It is a consequence of a deep theorem implying a difference between area and volume that is more fundamental than the obvious difference between two- and three-dimensions.) A radically different idea is needed to show the two volumes are equal.

The radical idea is to think of a three-dimensional shape as a stack of thin layers. For example, a rectangular box becomes a deck of cards.

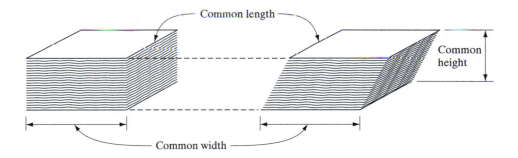

Then, if you push sideways on this deck (i.e., shear the rectangular box), you get a second solid that has the same volume. The second solid also has the same base and height as the original box.

Do the same thing with a triangular pyramid. Imagine that it is made of layers, like a triangular deck of cards. You can push sideways on this deck in many ways to make other triangular pyramids that have the same height and base and volume as the original. It seems reasonable to conclude that in this way you can get any triangular pyramid having the same height and base as the original.

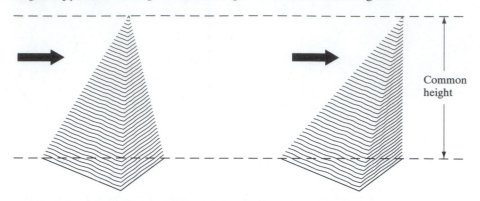

Common
height

From this we obtain the following theorem.

Theorem 1. Two triangular pyramids with the same base and height have the same volume.

Problem 28

The discussion preceding the theorem provides an informal argument for the plausibility of the theorem itself. The object of this problem is to make this argument more precise. We will use ideas from integral calculus as a guide.

(a) First we have to be precise about what we mean by "layer." We think: The volume of a pyramid is to be the sum of the volumes of the layers. The layers, as we were thinking about them above, didn't have thickness. Furthermore, there were infinitely many of them. How to deal with infinitely many layers, each of zero volume? Integral calculus handles it by making the layers have thickness, then letting the thickness go to zero in a limit. How do we get a handle on the layers that have thickness? Again, integral calculus has an answer: consider a line segment perpendicular to the base of the pyramid, running up from the plane of the base to a length equal to the pyramid's height. Divide this segment into a number of equal pieces each of width equal to $\triangle x$.

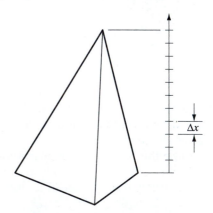

$\triangle x$

Through each point subdividing the segment pass a plane parallel to the plane of the pyramid's base. Include the endpoints of the segment as points of the subdivision. All these planes will slice up the pyramid. Define a layer to be the part of the pyramid between two planes. The pyramid is thus the union of all these layers.

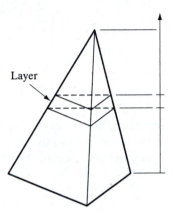

Layer

(b) Now each layer is really a shape like this:

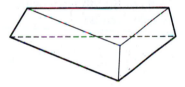

This is a truncated pyramid, a shape whose volume formula we would know if we knew that of a pyramid (which we don't). Another calculus idea is to approximate this thin layer by a shape whose volume we know: a triangular prism.

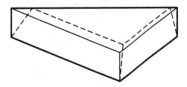

If we were to approximate all the layers by triangular prisms, the whole pyramid would be approximated by the following stair-step pyramid.

The volume of the stair-step pyramid is $A_1 \triangle x + A_2 \triangle x + \cdots A_n \triangle x$, where A_i is the cross-sectional area of the ith plane with the original pyramid. The volume of the original pyramid would then be

$$\lim_{n \to \infty} \sum_{i=1}^{n} A_i \triangle x$$

We can do the same thing with the other (tilted) pyramid, the one with base and height congruent to the original. Approximate its volume with a stair-step pyramid whose volume is $B_1 \triangle x + B_2 \triangle x + \cdots B_n \triangle x$, where B_i is the cross-sectional area of the ith plane with the tilted pyramid. The volume of the tilted pyramid would then be

$$\lim_{n \to \infty} \sum_{i=1}^{n} B_i \triangle x$$

Show that $A_i = B_i$, for $i = 1, 2, \ldots, n$ and complete the proof of the theorem.

Problem 29   Show that the volume of an oblique prism is also equal to the area of its base times its height. Use a "layered" argument as above.

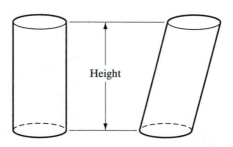

Height

Three triangular pyramids make up the triangular prism—the original and two others. Pyramids 1 and 2 have congruent bases and equal heights. Pyramids 2 and 3 have the same property, for different choice of base. Theorem 1 implies that all have the same volume. Thus the original is one-third the volume of the prism. Consequently, another theorem follows.

Theorem 2. The volume of a triangular pyramid is given by the following formula:

$$\text{volume} = (\tfrac{1}{3}) \cdot \text{area of base} \cdot \text{height}$$

What a nice formula! The analogous formula in two dimensions is for the area of a triangle:

$$\text{area} = (\tfrac{1}{2}) \cdot \text{base} \cdot \text{height}$$

Do the denominators of the two fractions which begin the formulas have anything to do with dimension?

Problem 30 Find a formula for the volume of a square-based pyramid.

Problem 31 A pattern is shown below. Make three copies. Cut out each one and assemble it. Here's a puzzle: the three assembled objects fit together to make a cube. Show how to do this. What does this say about the solution to the previous problem?

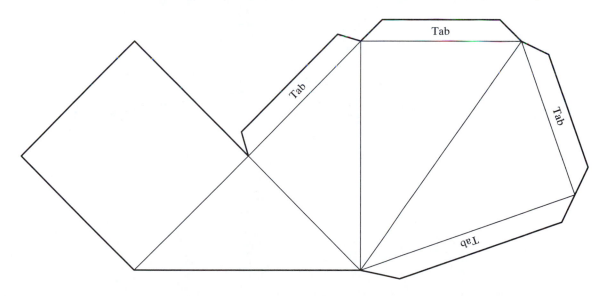

Problem 32   A pattern is shown below. Make six copies. Cut them out and assemble them. Here's a puzzle: the six objects fit together to make a cube. How can this happen? What does this say about the solution to problem 30?

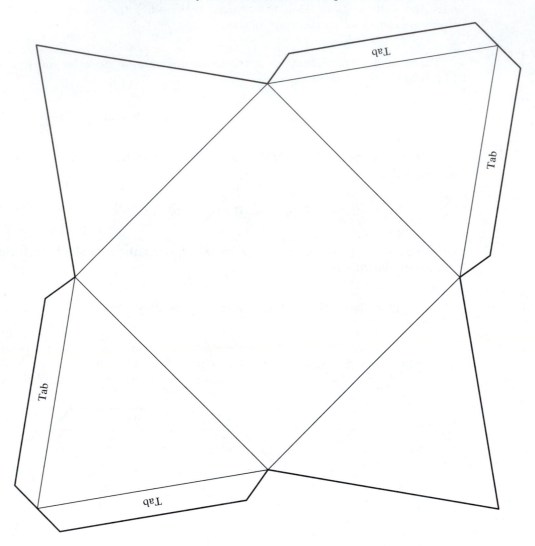

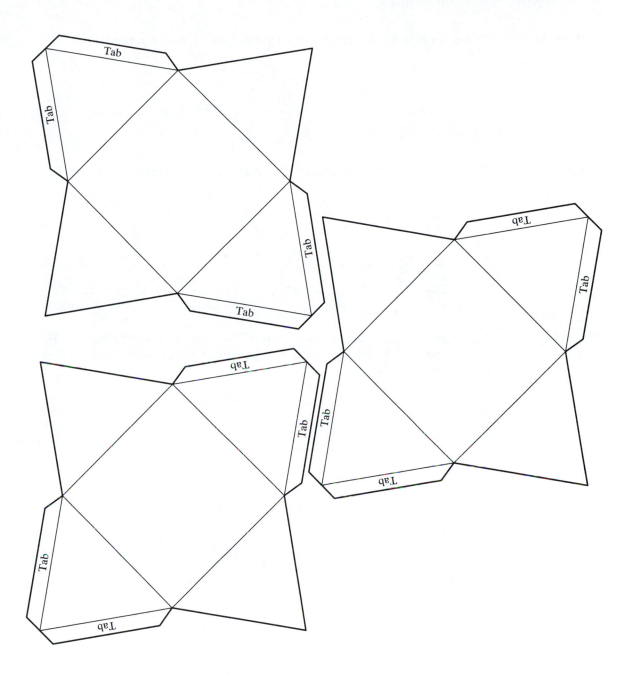

If you prefer a smaller template, duplicate this sketch twice to obtain the six objects you'll need to make a cube.

Problem 33   Find a formula for the volume of a pyramid whose base is a polygon.

Problem 34   Find a formula for the volume of a cone and give an argument to justify it.

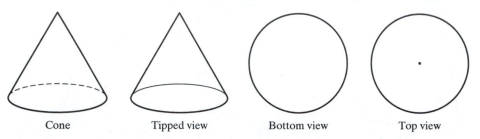

Cone   Tipped view   Bottom view   Top view

Problem 35   You have a conical-shaped pile of gravel. You want to know its volume. How do you find out? (Remember, you are responsible for making any linear measurements that are needed.)

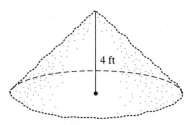

4 ft

Problem 36   When the rain gauge is full, it indicates that 1 in. of rain has fallen. It rained last night, and the gauge is filled up to the mark shown, which is halfway down the side of the gauge. How much rain fell last night?

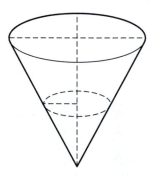

Problem 37   Find a formula for the volume of the frustrum of an Egyptian pyramid. (A frustrum of a pyramid is a pyramid whose top has been sliced off, parallel to the pyramid's base.)

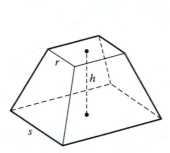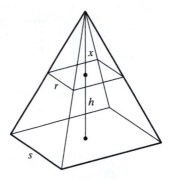

Problem 38   Find a formula for the volume of the frustrum of any old pyramid. What's a shape in the plane that's analogous to a frustrum of a pyramid? How do the corresponding formulas compare?

Problem 39   Find a formula for the volume of the frustrum of a cone.

Problem 40   Here is a tetrahedron. The lengths of its sides are not necessarily equal but two of its sides (of lengths r and s), when extended, do not intersect. The same two sides don't lie in the same plane, and their directions are perpendicular to each other. The shortest distance between them is h. Find the volume of this tetrahedron in two ways. One way to do it is to slice it into two smaller triangular pyramids as in the following diagram.

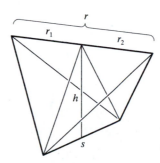

Another way to do it is to imbed it into a prism with a parallelogram base.

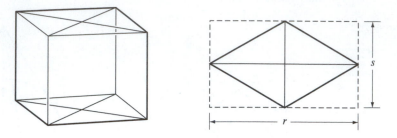

The tetrahedron will remain in the middle after slicing off four corners of the prism.

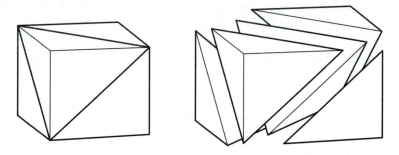

Problem 41

Find the volume of a square anti-prism. The top and bottom are congruent squares of side *s* lying in parallel planes, one square rotated 45° with respect to the other. The remaining sides are triangles as shown. The distance between the planes of the two squares is *h*. Find the volume of the anti-prism. Do it in the two ways suggested by the following sets of diagrams.

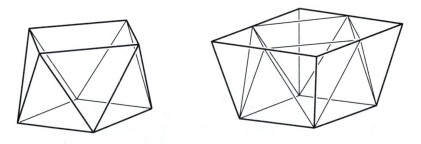

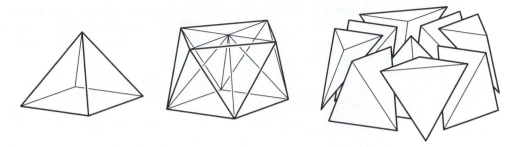

Problem 42   Do the methods of the previous problem generalize? For example, do they work when the two squares are of different size? Do the methods work for an anti-prism whose top and bottom are congruent regular pentagons. Do they work for an anti-prism whose top and bottom are congruent, nonsquare rectangles?

Problem 43   (Thanks to Thomas Banchoff in *Geometry's Future* for this problem.) Air France provides its customers a cup that is circular on the top and square on the bottom. The cup is convex. The top and bottom are in parallel planes. A sketch of the cup is shown below.

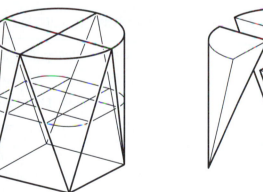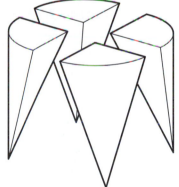

What is the volume of this cup? (The previous diagram suggests a possible dissection of the cup.) The airline provides a hole in the drop-down tray into which one can drop the cup so that it goes down halfway. What is the shape of the hole?

A Solution to the Sphere Volume Problem, Resumed. We were working on the volume of a sphere before we got sidetracked with finding the volume of a triangular pyramid. When we left off, we had dissected the sphere into little wedges which met in the center of the sphere.

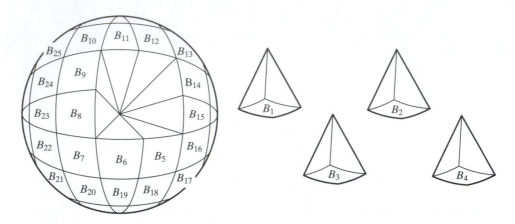

Each little wedge is approximately a pyramid. The volume of the sphere is the sum of the volumes of these little wedges. The volume of each wedge is approximately

$$\tfrac{1}{3} \cdot \text{area of base} \cdot \text{height}$$

But the height of each wedge is the radius R of the sphere. So we have the following:

$$\text{volume of sphere} = \text{sum of volumes of wedges}$$
$$\approx \tfrac{1}{3} \cdot R \cdot \text{area of base of wedge 1}$$
$$+ \tfrac{1}{3} \cdot R \cdot \text{area of base of wedge 2}$$
$$+ \tfrac{1}{3} \cdot R \cdot \text{area of base of wedge 3}$$
$$+ \cdots$$
$$= \tfrac{1}{3} \cdot R \cdot \text{sum of areas of bases of wedges}$$

Furthermore, the sum of the areas of the bases of all the wedges is equal to the surface area of the sphere. That's interesting! It seems reasonable, then, that a formula for the volume of the sphere is the following:

$$\text{volume of sphere} = \tfrac{1}{3} \cdot R \cdot \text{surface area of sphere}$$

We got to this point by exploiting the analogy between the circle and the sphere. In the analogy, the area of a circle corresponds to the volume of the sphere. Now let's compare the corresponding formulas.

$$\text{area of circle} = 1/2 \cdot R \cdot \text{circumference of circle}$$

$$\text{volume of sphere} = 1/3 \cdot R \cdot \text{surface area of sphere}$$

Of course! The circumference of the circle corresponds to the surface area of the sphere. The circumference of the circle measures the circle's "skin"; the surface area of the sphere measures the sphere's "skin." What could be nicer! Again, there is the occurrence of 1/2 and 1/3. What could it mean?

We're not done yet. The formula for the circle relates its area with its circumference. The formula for the sphere relates its volume with its surface area. When we first came across the formula for the circle, we knew something about its circumference. In fact, we had a formula for it. As for the sphere, we don't know its surface area. Rats, again. (Doesn't this remind you of an earlier situation where we dissected both the circle and sphere into wedges? For the circle the wedges were roughly triangles whose area we knew. For the sphere the wedges were roughly pyramids whose volume we didn't know.)

Problem 44   We discussed several methods for estimating the circumference of a circle: rolling the circle on a piece of paper . . . , putting a string around the circle's circumference . . . , approximating the circumference by a bunch of straight line segments . . . . What are some analogous methods for estimating the surface area of a sphere? (Remember, the methods don't have to be sophisticated.) What do you think of the methods?

✺ *Archimedes Solution to the Sphere Volume Problem*

The formula we have for the sphere reduces the problem of finding its volume to that of finding its surface area. Rather than directly tackling the problem of finding the sphere's surface area, let's find the volume of the sphere by another method. The method is due to Archimedes. It doesn't follow the intuitive route of the previous solution, based on the analogy with the area of the circle. However, the argument is ingenious and accessible.

To find the volume of a sphere, Archimedes compares it with the volumes of a certain cone and a certain cylinder. To do this he balances volumes at two ends of a lever arm. He uses the law of the lever, whose formulation is also due to Archimedes.

The Law of the Lever.   Two weights w_1 and w_2 balance on a lever arm with fulcrum at F if

$$w_1 \cdot s_1 = w_2 \cdot s_2$$

where the distance from w_1 to F is s_1 and the distance from w_2 to F is s_2.

In Archimedes's solution, volumes will play the role of the weights.

Here is the setup. Start with a sphere having diameter *HJ* and center *X*. Imagine a cone superimposed on this setting. It has apex at *H* and its base is a circle which lies in a plane through *J*, perpendicular to the line *HJ*. Thus, the cone's height is equal to the diameter of the sphere and the cone intersects the surface of the sphere in a circle *YZ*. Imagine a cylinder *LMNP* also superimposed on this setting. The cylinder has the same base as the cone and the same height *HJ*.

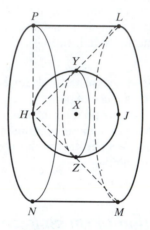

Extend the segment *HJ* an equal distance to *K*. Thus, *KH* = *HJ*. The line segment *KJ* is Archimedes's lever arm, with fulcrum at *H*. There are three solid shapes: the sphere, the cone, and the cylinder. Here is a diagram showing a cross section of the three shapes by a plane passing through the line *HJ*.

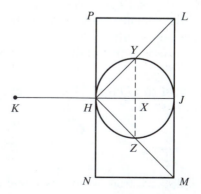

He will suspend two of the three shapes at point K to balance with the third solid suspended at a point along segment HJ. Archimedes's argument involves thinking of each solid as being made up of layers—or cross sections. In this case the cross sections are perpendicular to the line HJ. You recall that we used the layer idea in dealing earlier with triangular pyramids. Archimedes then suspends a solid at K (or wherever along the lever arm) one cross section at a time.

Take a typical point F on the line segment HJ. In the plane of the diagram, construct a perpendicular to HJ through F. This intersects the cone in D and C, the sphere in B and G, and the cylinder in A and E. The plane through F perpendicular to HJ intersects the three solids in cross sections, each of which is a circle. Points D and C are the end points of a diameter for the cone's circle; B and A are end points of a diameter of the sphere's circle; A and E are end points of a diameter of the cylinder's circle.

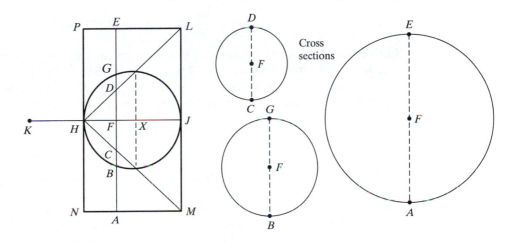

Then $FD = FH$ (HLM is an isosceles right triangle). Thus, by the Pythagorean theorem,

$$FD^2 + FG^2 = FH^2 + FG^2 = HG^2$$

Furthermore $HG^2 = HF \cdot HJ$.

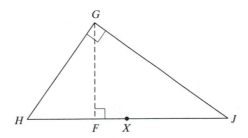

The angle at G is a right angle because triangle HGJ is inscribed in a semicircle with diameter HJ. By similar triangles, $HJ/GH = GH/HF$. Thus, also, $HJ \cdot HF = HG^2$.

Thus, $FD^2 + FG^2 = HF \cdot HJ$ so that $(FD^2 + FG^2) \cdot HJ = HF \cdot HJ^2$

But, since $HJ = FE$, this gives us

$$(FD^2 + FG^2) \cdot HJ = FE^2 \cdot HF,$$

and, since $HJ = HK$,

$$(FD^2 + FG^2) \cdot HK = FE^2 \cdot HF$$

Thus, multiplying both sides by π, we obtain

$$(\pi FD^2 + \pi FG^2) \cdot HK = \pi FE^2 \cdot HF$$

Let's interpret this last equation as if it were an instance of the law of the lever. The lever is KJ with fulcrum at H. The lengths of the two lever arms are HK and HF. Things are "weighed" (or suspended) at point K, on the one hand, and at point F, on the other. The things weighed are circles whose areas are as follows:

$\pi FE^2 =$ area of circle $AE =$ area of cross section of cylinder (at F)

$\pi FD^2 =$ area of circle $DC =$ area of cross section of cone (at F)

$\pi FG^2 =$ area of circle $BG =$ area of cross section of sphere (at F)

Thus, an interpretation of the equation $(\pi FD^2 + \pi FG^2) \cdot HK = \pi FE^2 \cdot HF$ in the context of the law of the lever says that if we suspend circles BG and DC at K they will balance the circle AE suspended where it is at F. In other words, if we suspend the cross section of the sphere and the cross section of the cone at K, it will balance with the cross section of the cylinder suspended at F (or just left alone where it is).

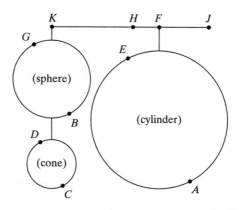

This is true no matter what point *F* we choose on the diameter *HJ*. Suspend at *K* all the circles (like the circle with diameter *CD*) which are cross sections (perpendicular to *HJ*) of the cone plus all the circles (like the circle with diameter *BG*) which are cross sections of the sphere. All of this will balance the cross sections (like the circle with diameter *EA*) of the cylinder hung just where they are.

Put another way, if we think of each shape as being made up of its cross sections, then the cone and sphere suspended at *K* will balance the cylinder sitting where it has been all along. Since the center of gravity of the cylinder is at *X*, the center of the sphere, we can suspend the cone and sphere at *K* and the cylinder at *X* and everything will balance:

$$(\text{volume of sphere} + \text{volume of cone}) \cdot HK = (\text{volume of cylinder}) \cdot HX$$

Since $HX = (\frac{1}{2})HK$, this gives us

$$\text{volume of sphere} + \text{volume of cone} = (\tfrac{1}{2}) \text{ volume of cylinder}$$

Finally, since we have formulas for the volumes of the cone and cylinder, this will give us a formula for the volume of a sphere.

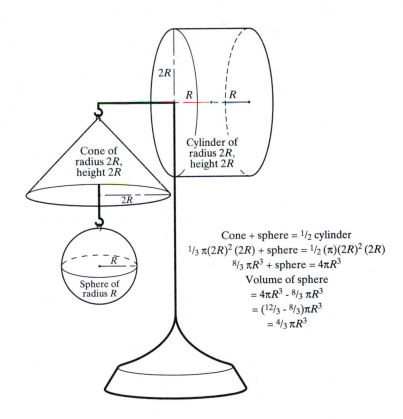

Cone of radius 2*R*, height 2*R*

Cylinder of radius 2*R*, height 2*R*

Sphere of radius *R*

Cone + sphere = $^1/_2$ cylinder
$^1/_3\, \pi(2R)^2\,(2R) + \text{sphere} = {}^1/_2\,(\pi)(2R)^2\,(2R)$
$^8/_3\, \pi R^3 + \text{sphere} = 4\pi R^3$
Volume of sphere
$= 4\pi R^3 - {}^8/_3\, \pi R^3$
$= (^{12}/_3 - {}^8/_3)\pi R^3$
$= {}^4/_3\, \pi R^3$

Problem 45   Using the equation above and formulas for the volume of a cone and the volume of a cylinder, obtain a formula for the volume of a sphere. Earlier, we obtained a formula relating the volume of a sphere with its surface area. Use the two formulas to find a formula for the surface area of a sphere.

Problem 46   Imagine a spherical egg of radius R. To eat what's inside the egg, you slice off a cap (of depth S along a radius). What is the volume of what's left? (What's left is what you eat, basically.) [Hint: Archimedes.]

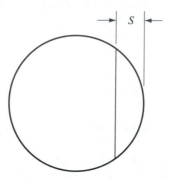

Problem 47   Archimedes (282–212 B.C.) lived in Syracuse, on the island of the Greek colony of Sicily. (Compare his dates with those of Plato, who flourished around 350 B.C., and Euclid, who flourished around 300 B.C.) When mathematics returned to Europe as a high art in the renaissance, the work of Archimedes was the ultimate model of good mathematics. With hindsight, one can also trace some basic concepts of the calculus back to Archimedes. You may see premonitions of calculus in Archimedes' argument for finding the volume of a sphere. Nevertheless, Archimedes felt that thinking of the volumes as being made up of infinitely thin layers kept his argument from being a good proof. Later, he established the relationship between the volumes of sphere, cone, and cylinder using a different argument, but one that he considered sufficiently "rigorous."

Archimedes found the informal argument involving infinitely thin layers useful for discovering the theorem but did not turn it into a rigorous proof. (This informal/heuristic argument is an instance of Archimedes' *Method*. See the Notes Section at the end of this chapter for a discussion of this.) We, however, have calculus. In problem 28 we saw how an argument using infinitely thin layers could be made more rigorous using ideas from the calculus. Do the same thing with Archimedes' informal argument: turn it into a rigorous one. Use the discussion in Problem 28 as a model. Here are some additional ideas getting the argument started: Select points on HJ which will divide the segment into n pieces of equal width $\triangle x$. Planes perpendicular to HJ and passing through these points will slice the three solids into thin layers of thickness $\triangle x$. Approximate these layers by cylinders of thickness $\triangle x$. Check that these thin cylinders "balance" just as the infinitely thin areas did:

$$(S_i \triangle x + Co_i \triangle x) \cdot HJ = Cy_i \triangle x \cdot i \cdot HJ/n$$

where S_i, Co_i, and Cy_i are the cross-sectional areas of sphere, cone, and cylinder by the *i*th plane ($i = 1, 2, \ldots, n$), the *n*th point being the point J. [Check this out. Make sure all the details are correct. This is just a sketch of the ideas!] What can you say about all the values $Cy_1, \ldots, Cy_n$?

Once again, the volume of the sphere will be the limit (as *n* gets large) of the sum of its approximating cylinders. Similarly for the volume of the cone. So, sum both sides of the equation above and take limits of both sides as *n* gets big. What happens? Fill in the details and clean up the argument.

Problem 48   A soap manufacturer makes a spherical-shaped soap that is packed in a box, 10 cm on a side, shown on the left. She is thinking of making eight smaller spherical soaps and packing them in the same-sized box. These are shown on the right. How does the volume of soap in the box on the right compare with the volume of soap in the box on the left?

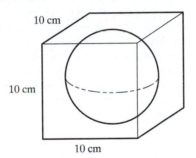
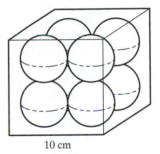

✳ *Full Circle*

In this chapter's trip through the familiar terrains of measuring length, area, and volume, there were three significant stopping-off points: the circumference of the circle, the area of the circle, and the volume and surface area of the sphere. We were able to come up with informal arguments making the traditional formulas for these quantities believable at a more basic level than the calculus. The idea was to provide new understanding of these formulas by connecting them with elementary ideas of geometry and measurement. We also wanted to show that finding these formulas was not an act of divine intervention but is humanly possible. In arriving at these formulas informally, we used approximation and analogy. We also showed, using limits and other calculus ideas, how we could turn these informal arguments into rigorous ones. (See the section entitled A More Convincing Argument, plus Problem 19, and Problem 47.) Along the way, we reviewed the development of the notion of area and, by analogy with area, the development of volume. Another stop on our trip also stood out: In dealing with the volume of a sphere, we came upon the problem of finding the volume of a pyramid. Considering the analogy of the volume of a pyramid with the area of a triangle led us to ask certain questions. To answer one of these questions we needed a very new idea.

∗ *Notes*

The roots of this chapter go way back in time. Measurement of length, area, and volume were at the beginnings of every civilization's first experiences with quantification. Many early written records consist of problems about land measurement and about quantities of grain sold, bought, or stored.

The oldest written documents describing uses of mathematics are around 4000 years old and come from the ancient Egyptian, Babylonian, and Chinese civilizations. From these sources it appears that the ancient peoples knew how to calculate the areas of triangles, rectangles, and trapezoids. There are many examples in the texts of the uses of $a = bh$ and $a = (1/2)bh$ for the area of a rectangle and triangle, respectively.

The circle created some difficulty, however. All these peoples seemed to know that the circumference of a circle is some fixed number times its radius and that the area is equal to some fixed number times its radius squared. The Babylonians and the Chinese knew that the two fixed numbers were related, probably using an argument similar to the one we used; their algorithms for calculating the area of a circle suggested our modern formula $A = (C/2)(d/2)$, C and d being the circumference and diameter, respectively. However, the Egyptians did not know that the two numbers were related.

The Babylonians appeared to use $C = 3d$, giving π the value of 3. As confirmation of this, the Hebrew Bible gives it this value in I Kings 7:23, a passage dealing with the reign of Solomon in 950 B.C.: "And he made a molten sea of ten cubits from brim to brim, round in compass . . . and a line of thirty cubits did compass it round about."

In the Rhind Papyrus (Egyptian, c. 1800 B.C.) the following problem occurs: "Example of a round field of diameter 9. What is the area? Take away 1/9 of the diameter; the remainder is 8. Multiply 8 times 8; it makes 64. Therefore, the area is 64." In other words, the area is given by $A = [(8/9)d]^2$. That makes for 3.16046 as a value for π.

In his treatise, On the measurement of the circle, Archimedes (c. 287–212 B.C.) offers the first recorded method for obtaining closer and closer approximations to the value of π. He showed that $3\ 1/7 > \pi > 3\ 10/71$ using basically the method described in problem 1.

A significant advance in calculating volumes is described in the Moscow Papyrus (Egyptian, c. 1800 B.C.) where the algorithm for finding the volume of a truncated pyramid, when translated into modern form, leads to the correct formula $V = (h/3)(a^2 + ab + b^2)$—$h$ is height, and a and b are sides of the square bases. In classical Greek times, Democratus (fifth century B.C.) knew that the volume of a pyramid was 1/3 the area of its base times its height. A proof by Eudoxus (c. 409–c. 356 B.C.) of this fact appears in Book XII of Euclid's Elements (c. 300 B.C.). Book XII also deals with the volumes of prisms, including the volume of a cylinder.

The layering argument we used in dealing with the volume of a pyramid is an instance of Cavalieri's Principle, due to the Italian mathematician Bonaventura Cavalieri (1598–1647). The Principle reads as follows.

If two shapes in space have equal altitudes and if cross sections made by planes parallel to their bases and at equal distances from them have their areas always equal, then the two shapes have equal volumes.

Cavalieri's argument justifying this principle builds on Archimedes' heuristic (set out in Archimedes' *Method*) in which he thinks of the volume of a three-dimensional shape as the "sum" of the areas of parallel planar cross sections. In his *Method,* Archimedes also thought of an area as being made up of parallel line sections of the shape and used the idea to calculate the area between a parabolic curve and a straight line. Galileo and Kepler used this kind of thinking in dealing with areas; Cavalieri has a version of his Principle for area as well as for volume. These ideas were eventually incorporated into calculus, created a few years after Cavalieri's death. After Newton and Leibnitz, the calculus became the most versatile tool for calculating areas and volumes. The notions of calculus were not put on the rigorous foundation that Archimedes might accept until the beginning the nineteenth century.

There are some interesting and significant twentieth-century contributions to the theory of volume. You recall that determining the volume of a pyramid involved a kind of reasoning not used in finding volumes of prisms and in finding areas of polygons. The proof in Euclid's *Elements* for finding the volume of a pyramid includes arguments reserved for finding the areas and volumes of shapes with curved edges or faces, such as circles, cylinders, and spheres. In the plane, to calculate the area of any polygon, you dissect it into triangles. The area of each triangle in turn can be calculated by complementing it with a congruent triangle (or pieces of the same) to make a rectangle, whose area is easily determined. Why can't something similar be done in space for shapes that are analogues to polygons—shapes bounded by pieces of planes, i.e., polyhedra? Why can't we calculate the volume of a polyhedron by dissection, complimentation, and eventually the calculation of the volumes of a bunch of rectangular boxes? This is essentially the question posed as Hilbert's third problem. In an address in 1900 to the International Congress of Mathematicians, the great German mathematician, David Hilbert, challenged the mathematical community, in the century to come, to solve a set of problems whose solution he felt to be the most important for the progress of twentieth-century mathematics. Since then many of these problems have been solved and the occasion of each solution celebrates a benchmark in the history of twentieth-century mathematics. (See the reference, *Hilbert's Problems* (1977), by Irving Kaplansky.) Unknown to Hilbert, the third problem had been solved by the German, Max Dehn, a few months before Hilbert's address. (See the reference by Boltnianskii (1963) below. See also Project C5 in Chapter 9, which deals with Dehn's solution.) The negative solution to Hilbert's third problem indicates one fundamental difference between the nature of two-dimensional space and that of three-dimensional space. Another difference is suggested by the famous paradox of Banach-Tarski (1924) in which a sphere of radius 1 in. can be dissected into a finite number of pieces which are then reassembled to form a sphere the size of the earth. Of course, the dissection must be pretty wild! [See the reference, *The World of Mathematics* by James Newman (1956), pp. 1944–5, for a summary of this result and *The Banach-Tarski Paradox,* by Stan Wagon (1985), in Volume 24 of the *Encyclopedia of Mathematics and its Applications*, for more details.]

✳ References

Banchoff, Thomas F., *Beyond the third dimension: geometry, computer graphics and higher dimensions.* New York: Scientific American Library, Distributed by W. H. Freeman, 1990.

Banchoff, Thomas F., *et al., Geometry's future.* Arlington, MA: COMAP, Inc., 1990.

Beck, A., Bleicher, M. N., and Crowe, D. W., *Excursions into mathematics.* New York: Worth Publishers, 1970.

Boltianskii, V. G., *Equivalent and equidecomposable figures.* Boston: Heath, 1963.

Heath, Thomas Little, *The thirteen books of Euclid's Elements.* New York: Dover, 1956.

Heath, Thomas Little, *The works of Archimedes.* New York: Dover, 1953.

Kaplansky, Irving. *Hilbert's problems.* Lecture Notes in Mathematics. Department of Mathematics, University of Chicago, 1977.

Katz, Victor, *A history of mathematics: an introduction.* New York: Harper Collins, 1993.

Lines, L., *Solid geometry.* New York: Dover, 1965.

Newman, James. *The world of mathematics.* New York: Simon and Schuster, 1956.

Wagon, Stan, *The Banach-Tarski paradox.* Vol. 24. In *Encyclopedia of Mathematics and its Applications.* New York: Cambridge University Press, 1985.

3 ☀ *Polyhedra*

In the previous chapter, there were several occasions where experience with (two-dimensional) area gave us ideas for dealing with (three-dimensional) volume. We established an analogy between area and volume that connected shapes, processes, and principles. Sometimes the analogy was very clean: the two systems seemed to be identical; there was smooth sailing in utilizing the one to solve problems in the other. At other times the analogy seemed to break down: the two systems became quite different; at these times we were forced to be creative and come up with new ideas. We were amazed at the similarities **and** the differences. The analogy energized us.

We want to do something similar in this chapter. This time we want to set up an analogy between polygons, two-dimensional objects we have a lot of experience with, and their three-dimensional counterparts, whatever they might be. We will want to "figure out" what characteristics these three-dimensional objects should have, i.e., we will want to come up with a loose definition. Two things will guide us: first, we will briefly analyze polygons in order to come up with ingredients that might have three-dimensional counterparts; second, we will take an inventory of familiar three-dimensional objects that "feel" like three-dimensional relatives of two-dimensional polygons. After we are satisfied with our working definition of "3-D analogue to a 2-D polygon," we will try to find objects that satisfy this definition and acquire some understanding of them. This will be an exploration of unknown territory for most of us. But again, the analogy will guide us. Things we know about polygons in the plane will suggest directions in our explorations. For example, one of our favorite collections of polygons is the regular polygons. What characteristics should the three-dimensional counterparts to regular polygons have? What are some three-dimensional objects that have these characteristics? What do they look like? As we explore, we anticipate easy going sometimes; and we expect there will be surprises (even difficulties) at others. There will be new questions, too, that we never thought to ask of polygons in the plane.

☀ What Is a Three-Dimensional Counterpart to a Polygon in the Plane?

We know what a polygon in two-dimensions is. What is its counterpart in space?

To try to answer this, let's list some characteristics of a polygon in the plane. It has edges and vertices. The vertices are the end points of the edges. Each vertex joins two edges of the polygon. The polygon has an interior. The interior is completely enclosed by the edges. Furthermore, the collection of edges and vertices is connected: you can travel from one point on an edge to another by a path through the edges.

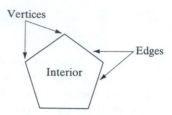

There are already three-dimensional shapes we think of as counterparts to specific polygons. A rectangular box is the three-dimensional counterpart to a rectangle. A triangular pyramid is the three-dimensional counterpart to a triangle.

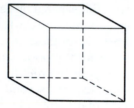 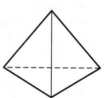

Once having mentioned these shapes, we are reminded of others: a triangular prism, an Egyptian pyramid.

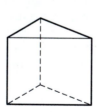 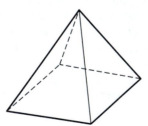

Our feeling is that, whatever general definition we come up with for a three-dimensional counterpart to a plane polygon, it ought to include all of these shapes. What are some of *their* characteristics? Each shape has flat faces, each of which is a polygon. Each shape has edges. These edges have endpoints (vertices). Each shape has an interior.

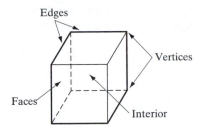

We make the following comparison of these ingredients:

Ingredients

Polygon	*3-D Shape*
Vertices (zero-dimensional objects)	Vertices (zero-dimensional objects)
Edges (one-dimensional objects)	Edges (one-dimensional objects)
Interior (two-dimensional object)	Faces (two-dimensional objects)
	Interior (three-dimensional object)

Next, let's compare how these ingredients relate.

Relationships between Ingredients

Polygon	*3-D Shape*
Each vertex belongs to exactly two edges	Each edge belongs to exactly two faces
Edges meet only at vertices	Edges meet only at vertices
Each vertex is a common meeting point for two edges	Each vertex is a common meeting point for three or more edges
Interior completely enclosed by the edges	Interior completely enclosed by the faces
The set of edges and vertices is connected	The set of faces, edges, and vertices is connected

The lists of ingredients and the relationships between them are similar enough to give us a definition of a three-dimensional counterpart to a two-dimensional polygon. We will call the three-dimensional shape a **polyhedron** (Greek for "many faced") and define it as being a three-dimensional object with ingredients and relationships as listed under "3-D shape" above. This seems to be a good starting point. It may not pinpoint completely what we are looking for. (We are not certain of exactly *what* we are looking for.) It should serve us well until we have to change it!

Now that we have a working definition, let's do some exploring. The following problems will get our feet wet and give us some experience with polyhedra (plural of polyhedron).

Problem 1   A polygon in the plane is **convex** when the line segment joining any two points in the polygon (= interior plus edges and vertices) itself lies entirely in the polygon.

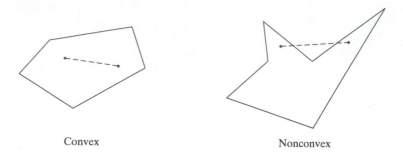

Convex   Nonconvex

Can you think of a nonconvex polyhedron?

Problem 2   One familiar polyhedron is the cube which can be cut open along edges and flattened out into a sort of "dressmaker's pattern." Here is a dressmaker's pattern for a cube.

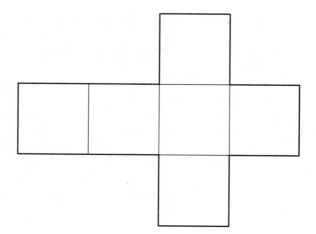

Cut this out, fold (away from you) along the solid lines, then assemble to make a nice cube. (Scotch tape may help.)

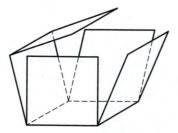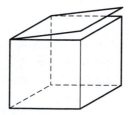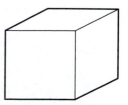

Problem 3   You notice that the dressmaker's pattern for a cube is made of up six squares. Which of the following planar configurations are dressmaker's patterns for a cube?

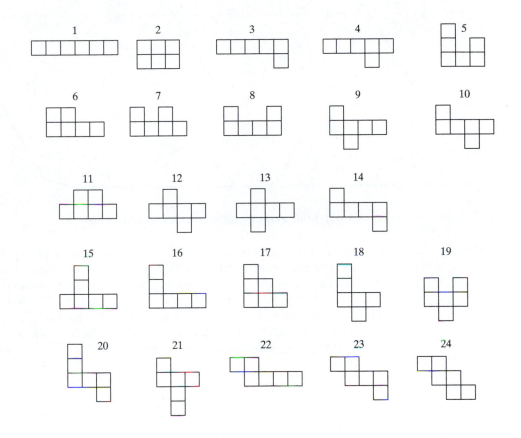

Problem 4   The faces of the cube in the picture below have been numbered from 1 through 6. Select six of the dressmaker's patterns in the previous problem that make a cube and label the squares with numerals 1 through 6 in such a way that, when assembled, the labels will appear just as in the picture. (Extra credit goes to those who can do this so that the numerals are all oriented correctly!)

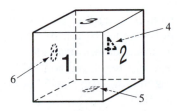

Problem 5   Here is a dressmaker's pattern for the triangular pyramid, each of whose faces is an equilateral triangle.

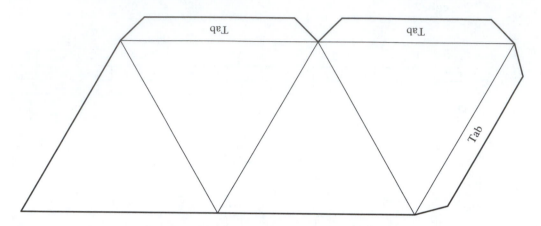

Cut this out, fold, and assemble into the polyhedron.

Problem 6   An easily described class of polyhedra consists of pyramids with regular polygons for bases. You are already familiar with a pyramid having an equilateral triangle as base and a pyramid having a square as base (an Egyptian pyramid). For every $n > 2$, there is a pyramid whose base is a regular n-gon. This is a polyhedron with $n + 1$ faces, one of which is a regular n-gon (the base of the pyramid) and n of which are triangles attached to each of the n-gon's edges. The n triangles meet at a point (the pyramid's apex). A pyramid whose base is a regular pentagon is shown below.

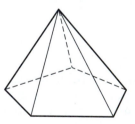

Design an accurate dressmaker's pattern for this shape, make a copy of it, and assemble it into the polyhedron.

Problem 7   Another easily described class of polyhedra consists of prisms with different regular polygons for bases. A cube is a square-based prism. A triangular prism may have an equilateral triangle as its base. For every $n > 2$, there is a prism whose base is a regular n-gon. This is a polyhedron consisting of two parallel regular n-gons (its "top" and "bottom") joined by rectangles (n of them). A prism whose base is a regular hexagon and whose sides are squares is shown below. Design a careful dressmaker's pattern for this, make a copy of it, and assemble.

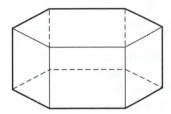

Problem 8   Yet another class of polyhedra consists of the **anti-prisms**. For every $n > 2$, there is a polyhedron consisting of two parallel regular n-gons (its "top" and "bottom") as in a prism whose base is a regular n-gon. This time one n-gon has been given a twist (a rotation of π/n about its center), and the two n-gons are connected by equilateral triangles. An anti-prism with regular hexagonal base is shown below. Design a careful dressmaker's pattern for this. Make a copy of it and assemble.

[You can view this anti-prism in The Geometry Files. A good exercise to get acquainted with this software is to use it to construct an anti-prism with a regular octagonal base.]

Problem 9   We classify polygons crudely by the number of edges they have. A triangle is a polygon with three edges; a quadrilateral is one with four edges; an n-gon is one with n edges. A way to classify polyhedra might be by their number of faces. For example, a cube has six faces and it is sometimes called a *hexahedron*. An Egyptian pyramid has five faces, so we might call it a 5-hedron (or pentahedron). We know that for every $n > 2$, there is a polygon with n sides—an n-gon. Is there such a thing as a 7-hedron? That is, is there a polyhedron having exactly seven faces? For which n is there an n-hedron?

Problem 10   Given a polyhedron, we can obtain another polyhedron by cutting off a corner where a vertex is. This process is called **truncation**. For example, if we cut off the corners of a cube, we obtain a shape that looks something like the following.

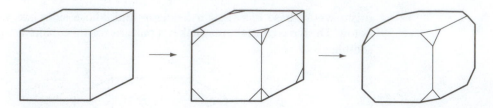

If we make the cuts so that the faces of the resulting shape are equilateral triangles and regular octagons, then we get a shape called the **truncated cube**. Show that all the edges of a truncated cube have the same length. Find the volume of a truncated cube all of whose edges have the common length of s. [A good thing to do would be to construct the truncated cube with The Geometry Files.]

✳ *Regular Polyhedra*

A polyhedron is a space counterpart to a polygon in the plane. What is a counterpart in space to a **regular** polygon in the plane? Any pair of edges of a regular polygon is congruent, that is, all the edges have the same length. Any pair of vertex angles is congruent.

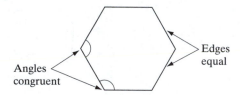

Angles congruent

Edges equal

Let's see if we can come up with a definition of **regular polyhedron**, the analogue of a regular polygon. Imagine what such a polyhedron would be. Certainly its polygonal faces should be congruent, all its edges should have the same length, and any pair of vertex angles on the faces should be congruent. These three conditions imply that the faces of such a polyhedron are all regular polygons. (That's nice! We must be on the right track.) Since all the regular polygonal faces are congruent, there is a number, n, attached to the polyhedron. Every face of the polyhedron is a regular n-gon. Do we know any polyhedra which satisfy our conditions? Two examples which come to mind are a cube with square faces ($n = 4$) and a triangular pyramid with equilateral triangular faces ($n = 3$).

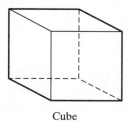

Cube

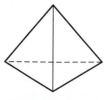

Triangular pyramid

Are there others? Consider the following polyhedron formed by taking two copies of the triangular pyramid above and gluing them together along a common face to form a double pyramid:

Triangular double pyramid

Do we want to call this shape a regular polyhedron? (Remember, we—you and me—are in charge of making definitions. If we don't like what we have, we can change it.) It seems to satisfy all of our conditions. Let's have a close look at it. There's a vertex where three faces meet. There's another vertex where four faces meet. For the cube, three squares meet at each vertex. For the triangular pyramid, three triangles meet at each vertex. For the cube, any pair of vertices "looks the same"; for the triangular pyramid, it's the same thing. Going back to a regular polygon in the plane, any pair of vertices "looks the same." The double pyramid lacks the grace and symmetry that the cube and triangular pyramid have. Let's add a new condition to our definition of regular polyhedron: any two vertices must "look the same."

What does "look the same" mean? Whatever it means, having the condition be satisfied surely implies that the number of faces meeting at one vertex is the same as the number of faces meeting at any other vertex.

Problem 11   Make a precise definition of what it means for two vertices of a polyhedron to "look the same."

Problem 12   We normally think of a regular polygon as being convex. Can you think of a polyhedron that satisfies the latest conditions for a regular polyhedron but that is not convex?

Problem 13   Think of a polyhedron having the following properties: all faces congruent, all vertices "look the same," but not all edges equal. Does such a polyhedron exist?

Problem 14   Think of a polyhedron having the following properties: all faces congruent, all vertices "look the same," all edges equal, but the faces are not regular polygons. Does such a polyhedron exist?

✳ *Classification of Regular Polyhedra*

Suppose that the number of faces meeting at each vertex of a regular polyhedron is m. This gives us two numbers attached to the regular polyhedron: n, the number of edges of each and every face and m, the number of polygons meeting at each and every vertex. For the cube, $n = 4$, $m = 3$; for the triangular pyramid (= **regular tetrahedron**), $n = m = 3$.

For every natural number $n \geq 3$, there is exactly one (convex) regular n-gon. (We are being casual here. We are lumping all the squares into one pile, all the equilateral triangles into one pile, etc.) This is a complete list of regular polygons in the plane. It would be nice to have a complete list of (convex) regular polyhedra in space. We know that any regular polyhedron has a pair (n,m) of numbers attached to it where n and m have the meaning given above. Turn this around. Given a pair (n,m) of natural numbers, is there a regular polyhedron it's attached to? For example, is there a regular polyhedron all of whose faces are regular 11-gons with seven of these meeting at each vertex? The pair (11,7) would be attached to such a polyhedron. You think there is? not?

Suppose we have a regular polyhedron to which is attached the pair (n,m). Can we say anything about n and m? Are there any restrictions on n and m?

One thing we know is that $n \geq 3$. (Why is this?) We also know that $m \geq 3$. (Why is this?)

Before reading on it would be a good idea to try making a regular polyhedron. Start with a bunch of equilateral triangles as faces. Try various possibilities for m. The first job is to fit triangles around a vertex. Fiddle a bit. What do you find out?

Here's a similar problem. When we made a tessellation of the plane using regular polygons, we wanted several copies of a regular n-gon to fit around a vertex exactly—in the plane. (See Problem 29 of Chapter 1). This happens when some integer multiple (m) of the vertex angle of the n-gon equals 2π.

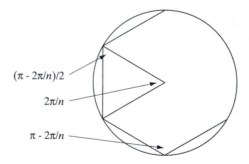

$(\pi - 2\pi/n)/2$

$2\pi/n$

$\pi - 2\pi/n$

In other words, $m(\pi - 2\pi/n) = 2\pi$. Or, $m(1 - 2/n) = 2$. Or,

$$1/m + 1/n = 1/2$$

The problem for a polyhedron is slightly different. We don't want the regular n-gons that fit together at a vertex to lie in a plane. But they do have to fit together in space. If the polyhedron is to be convex, then m times the vertex angle of a regular n-gon must be **less than** 2π. That is,

$$m(\pi - 2\pi/n) < 2\pi$$

or

$$m(1 - 2/n) < 2$$

or

$$1/2 < 1/m + 1/n \qquad (*)$$

For the *equation* $1/2 = 1/m + 1/n$, there were only a few solutions (n,m). For the *inequality* $1/2 < 1/m + 1/n$, it might seem that there are infinitely many solutions (n,m). Let's see. We know that $3 \leq m$, so that $1/m \leq 1/3$. Thus,

$$1/2 < 1/m + 1/n \leq 1/3 + 1/n$$

or

$$1/2 < 1/3 + 1/n$$

or

$$1/6 < 1/n$$

or

$$n < 6$$

We also know that $3 \leq n$. Using the argument just given (the roles of n and m in the inequality $(*)$ are the same), we find that also $m < 6$.

Thus, if a pair (n,m) is associated with a regular polyhedron, such a pair must satisfy

$$3 \leq n < 6 \quad \text{and} \quad 3 \leq m < 6.$$

There are only nine possible pairs! Here they are:

$$(3,3), \quad (3,4), \quad (3,5),$$
$$(4,3), \quad (4,4), \quad (4,5),$$
$$(5,3), \quad (5,4), \quad (5,5).$$

That's really extraordinary. Not infinitely many pairs! But nine pairs!

Problem for Right Now! Do all of the nine pairs satisfy the inequality (*)? Try them and see before reading on.

✳ Possible Regular Polyhedra

Theorem 1. Only five of the nine possible pairs associated with a regular polyhedron satisfy inequality (*). The five are given by the following table.

n	m
3	3
3	4
3	5
4	3
5	3

This list certainly narrows the possibilities somewhat. The only faces possible for a regular polyhedron are equilateral triangles, squares, and regular pentagons!

Problem for Right Now! But do the possible pairs correspond to regular polyhedra? Given a pair on the list, is there a regular polyhedron attached to it? Before reading on, take out bunches of equilateral triangles, squares, and regular pentagons and see if you can make a polyhedron for each pair. *Googleplex* would be good to use for this exercise. Or *Polydron*, or *Geo-Rings* . . .

Problem for Right Now! We know right off that some items on the list correspond to real live, regular polyhedra. The pair (4,3) corresponds to the cube and (3,3) to the regular tetrahedron. Are there two or more really different regular polyhedra corresponding to the pair (4,3)? Are there two or more really different regular polyhedra corresponding to the pair (3,3)?

We'll return to these problems after a look at more properties of polyhedra in general.

✳ Numerical Data Associated with a Polyhedron

We classify polygons crudely by their number of edges. So given a polygon, we count the number of sides it has to tell us where it fits in this classification. A polygon has another set of features we could count: it has vertices. We think that this could offer us another crude classification of polygons. But, of course, we know better: the number of vertices of a polygon is equal to the number of its edges.

What about things to count on a polyhedron? Take a cube, for example. Just as for a polygon, we could count edges: a cube has 12 edges. We could also count vertices: a cube has eight vertices. Hmm. The number of edges is not the same as the number of vertices. Polyhedra are different from polygons! Something else we could count is faces: a cube has six faces. If we were classifying polyhedra according to these three numbers, then a cube might be called a 12-8-6-hedron.

That's interesting. Is there anything we can say about these numbers? Let's take some polyhedra we know, gather some data, and construct a table. For each polyhedron, V will denote the number of its vertices, E the number its edges, and F the number of its faces.

Polyhedron	*V*	*E*	*F*
Cube	8	12	6
Tetrahedron	4	6	4
Egyptian pyramid	5	8	5
Triangular prism	6	9	5

Can you say anything interesting about the numbers in this chart?

Problem for Right Now!

One thing to look for is a relationship among V, E, and F for a given polyhedron. Here are two ideas you might consider:

1. Polygons in the plane are analogous to polyhedra in space. A polygon in the plane has V vertices and E edges. What is a relationship between V and E?
2. Calculate V, E, and F for other polyhedra. Add what you get to the table above. For example, take the double pyramid we looked at earlier. Also, use the examples of polyhedra from problems 6, 7, and 8. These will give you several **infinite** families of examples to add to the table.

Now stare at the expanded table. Can you say anything now?

Problem 15

A really crude way to classify polyhedra would be by their number of vertices. For example, a cube has $V = 8$. For which n is there a polyhedron which has $V = n$?

Problem 16

Another crude way to classify polyhedra would be by their number of edges. For example, a cube has $E = 12$. For which n is there a polyhedron which has $E = n$?

Problem 17

If you ignore length of edges and sizes of vertex angles, then the classification of polygons by the number of edges is good. Suppose you do the same thing for polyhedra: ignore the length of edges, the area of faces, and the sizes of angles at the vertices of faces. Would the classification of polyhedra by the triple of numbers (V, E, F) be good? (One way to get into this is to try to find a pair of polyhedra that have the same *V*, *E*, *F* data but that you think are significantly different, i.e., you can't get one from the other by shrinking or stretching.)

✳ *Euler's Formula*

You may have noticed that, for all the polyhedra listed in the table in the previous section, the relationship

$$V - E + F = 2$$

holds. This is called **Euler's Formula**. We will prove it true for all polyhedra in the following.

Theorem 2.   If a (convex) polyhedron has *V* vertices, *E* edges, and *F* faces, then

$$V - E + F = 2.$$

Proof: Take a polyhedron. Imagine that it is made of rubber. Cut out one of its faces and flatten out the rest without tearing it. You should get a planar shape that looks like the following.

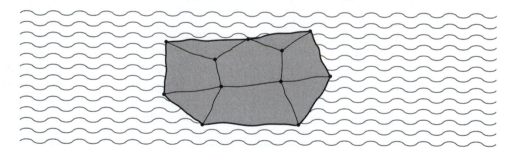

Imagine this as an island partitioned into countries (= polygons) with borders (= edges) and surrounded by a sea of water (= the removed face of the original polyhedron). For example, a flattened out version of the cube might look like the following.

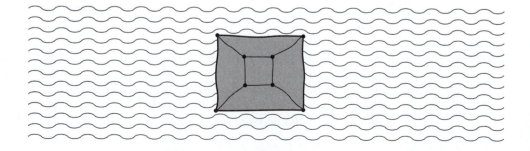

In the flattening out process, faces of the polyhedron become polygonal regions (countries) on the island, polyhedral edges become the edges (borders) of these polygons, and polyhedral vertices become the vertices of these polygons. There will be distortion. Lengths and areas will change during the flattening out operation. Straight lines may become curved. What should be preserved, however, are the quantities V, E, and F—except for the one face that was removed from the original polyhedron. We will have proved the theorem if we can show that $V - E + F = 1$ for the "island" divided into countries.

The strategy for showing $V - E + F = 1$ is to alter the island systematically in stages until it becomes one whose partitioning is simple, with vertices, edges, and faces that are easy to count. At each stage, the numbers V, E, and F will change, but we will show that $V - E + F$ will not.

Start with the given island having V vertices, E edges, and F faces. Remove a face of the island bordering along the water. At the same time we must also remove the edge

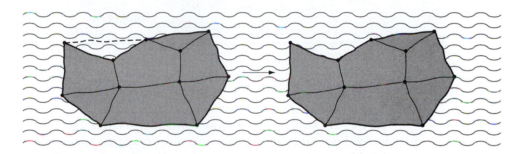

between the removed face and the water. Let V', E', F' be the number of vertices, edges, and faces, respectively, of the altered island. It's easy to see that

$$V' = V, E' = E - 1, F' = F - 1$$

Thus

$$V' - E' + F' = V - (E - 1) + (F - 1) = V - E + F$$

We summarize this in the following table.

	Vertices	Edges	Changes in Faces	Vertices − Edges + Faces
Stage #1	0	−1	−1	0

You will notice in the altered island that there is now one vertex or more attached to exactly two edges. These are unnecessary. We remove these one by

one. When we

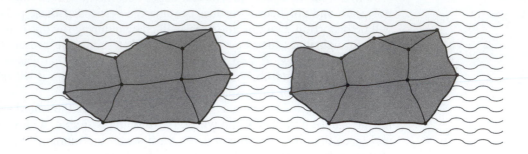

remove one (we call this stage 2), the number of vertices decreases by one and the number of edges decreases by one as well. Let's put this information in our table.

	Vertices	Edges	Changes in Faces	Vertices − Edges + Faces
Stage #1	0	−1	−1	0
Stage #2	−1	−1	0	0

Again, the number in the last column doesn't change!

After removing these "excess" vertices and edges, then, as before, we remove a face bordering on the perimeter of the altered island. (Make sure the face doesn't split the island into two pieces!) Again, the quantity in the last column doesn't change. We continue to remove a face bordering the water, then remove unnecessary edges and vertices. At each step, the quantity in the last column doesn't change. Eventually, we will arrive at the following.

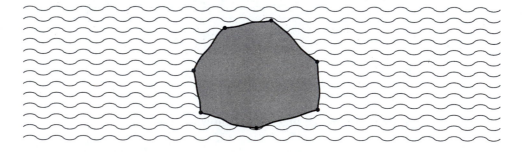

This is an island with a single polygonal face (country). The number of faces is 1, and the number of vertices is equal to the number of edges. For this simple island, the quantity

$$\# \text{ vertices} - \# \text{ edges} + \# \text{ faces}$$

is equal to 1. Since, at each stage in altering the original island, there was no change in the quantity # vertices − # edges + # faces, the number we get at the end of the process must be equal to the number we started with. Thus, $V − E + F = 1$ for the original island. ✳

Problem 18   The final polygon/island in the proof of the theorem has unnecessary vertices and edges. What would have happened if these had been removed before we counted them as above? What's going on?

Problem 19   In an earlier problem, we classified polyhedra crudely by their number of faces. A more refined classification scheme for polyhedra than number of faces is classification by **combinatorial type**. Two polyhedra are of the same combinatorial type if they have the same network of edges and vertices. Here are four polyhedra of the same combinatorial type:

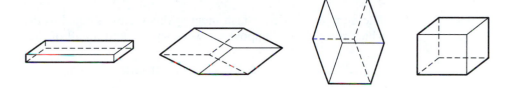

Here are three octahedra, each of a different combinatorial type:

How many combinatorial types are there for polyhedra having five faces? What are they? (Construct them!) Prove that these are all there are.

✳ *An Application of Euler's Formula*

The following theorem is an example of how Euler's formula can be used to obtain other information about a polyhedron.

Theorem 3.   Any polyhedron must have at least one face with five or fewer edges.

Proof: First of all, we assume there must be at least three faces which meet at a vertex. (Why do we assume this?)

Let V, E, and F denote the number of vertices, edges, and faces, respectively, of the polyhedron. Euler's formula is

$$V - E + F = 2$$

Next we want to find a relationship between V and E alone by using the polyhedron's vertices to count of the number of its edges. Each vertex of the polyhedron looks like the following.

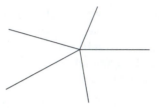

The number of edges which meet at this vertex is equal to the number of faces which meet there. Suppose we were able to add up all of these edges:

edges meeting at vertex #1

$+$ # edges meeting at vertex #2

$+$ # edges meeting at vertex #3

$+ \cdots$

What could this sum be equal to? Let's look at it in two different ways. From one point of view, each edge of the polyhedron has been counted twice in the sum, once for each of its end points. Thus, the sum above is equal to $2E$. From another point of view, each of the summands in the sum is at least three—since every vertex has at least three edges attached to it. And, the number of summands is equal to V. Thus, the sum is at least $3V$. Putting these two points of view together, we get

$$3V \leq 2E$$

or

$$V \leq (2/3)E$$

Let's combine this inequality with Euler's formula to obtain an inequality involving E and F:

$$2 = V - E + F \leq (2/3)E - E + F$$

or

$$2 \leq -(1/3)E + F$$

or, after some algebraic manipulation,

$$E + 6 \leq 3F \qquad (1)$$

The latter inequality expresses a relationship between E and F. Let's find another relationship. Earlier we counted edges by summing the edges attached to all the vertices. We could do something similar for faces: we know that each polygonal face has edges. Thus, consider the sum

$$\text{\# edges belonging to face \#1}$$
$$+ \text{ \# edges belonging to face \#2}$$
$$+ \text{ \# edges belonging to face \#3}$$
$$+ \cdots$$

Just as for the sum we wrote earlier (which counted edges using vertices), let's look at the present sum in two different ways. From one point of view, each edge of the polyhedron has been counted twice in the sum, once for each polygon it's an edge of. Thus, the sum above is equal to $2E$.

From another point of view, let F_i denote the number of faces having exactly i edges. Thus, F_3 is equal to the number of triangular faces the polyhedron has, F_4 is the number of quadrilateral faces it has, and so on. Now iF_i is the number of edges in the sum above contributed by the totality of all faces having i edges. Thus, by rearranging the summands of the sum above, we get $3F_3 + 4F_4 + 5F_5 + \cdots$.

Consequently, putting the two points of view together (once again), we have

$$2E = 3F_3 + 4F_4 + 5F_5 + \cdots \qquad (2)$$

Now, also

$$F = F_3 + F_4 + F_5 + \cdots \qquad (3)$$

Putting inequality (1) and equations (2) and (3) together, we obtain

$$12 + 2E \leq 6F$$

or

$$12 + 3F_3 + 4F_4 + 5F_5 + \cdots \leq 6(F_3 + F_4 + F_5 + \cdots)$$

or

$$12 + 3F_3 + 4F_4 + 5F_5 + 6F_6 + 7F_7 + \cdots \leq 6F_3 + 6F_4 + 6F_5 + 6F_6 + 6F_7 + \cdots$$

or

$$12 + F_7 + 2F_8 + 3F_9 + \cdots \leq 3F_3 + 2F_4 + F_5$$

The left-hand side of this inequality is positive—it's at least 12. Thus, the right-hand side of the inequality must also be positive. This implies that at least one of F_3, F_4, and F_5 must also be positive. This says that the polyhedron must have at least one face with three, four, or five edges. This proves the theorem. ✳

Problem 20   A quicker proof of this theorem might go something like this. You fill in the details. The proof is by contradiction: assume that all faces are polygons with six or more edges. Thus, $6F \leq 2E$. (Show this!) Since also $3V \leq 2E$ (as in the proof above), we can use Euler's formula to conclude that $12 \leq 0$. (Show this!) This is the contradiction that proves the theorem.

Problem 21   Suppose that a certain convex polyhedron has no triangular faces and no quadrilateral faces. What can you say about the number of pentagonal faces it has?

Problem 22   They say that, if a polyhedron has five faces, at least two of them must be triangles. Is this an old-wives tale, or what?

Problem 23   Suppose that a certain convex polyhedron has no pentagonal faces. What can you say about the number of triangular and quadrilateral faces that it has?

Problem 24   We mentioned the combinatorial type of a polyhedron earlier. I've been told that there are seven combinatorial types for polyhedra having six faces. Is this true? What are they? Proof?

Problem 25   Investigate all polyhedra having 12 or fewer faces, each of which is a quadrilateral.

Problem 26   Investigate all polyhedra having eight or fewer faces, each of which is an equilateral triangle.

✳ *Predicting with Euler's Formula*

Let's return to our investigation of regular polyhedra after having paused to rediscover and prove Euler's formula. Recall that a regular polyhedron is associated with a pair (n,m), where each face of the polyhedron is a regular n-gon and exactly m faces meet at each vertex. We obtained the following table of possibilities:

n	m
3	3
3	4
3	5
4	3
5	3

We know that the regular tetrahedron and cube are associated with the pairs (3,3) and (4,3), respectively. Are there polyhedra associated with the other three? Can more than one polyhedron be associated with a single pair?

Take a regular polyhedron of type (n,m). Let's use the values of n and m and Euler's formula to see if we can predict the values of V, E, and F for the polyhedron.

Let's count edges in two ways. (The idea for this analysis comes from the proof of the theorem in the last section.) On the one hand, every face is a regular n-gon so that every face has n edges. An addition of all edges for all faces yields nF. In this count, every edge is counted twice, once for each of the two faces it is an edge of. Hence

$$nF = 2E$$

On the other hand, m faces meet at each vertex. An addition of all edges for all vertices yields mV. In this count, every edge is counted twice. Hence

$$mV = 2E$$

These two equations imply that we can express each of V, E, and F in terms of E alone.

$$E = E, V = 2E/m, \qquad F = 2E/n$$

Thus

$$2 = V - E + F = 2E/m - E + 2E/n = (2/m - 1 + 2/n)E$$

or

$$E = 2(2/m - 1 + 2/n)^{-1}$$

We can solve for E exactly in terms of n and m; the same is true for V and F. Once you know n and m, then the values of E, V, and F are uniquely determined. Plugging n and m into these formulas, we get the following table.

n	m	E	V	F
3	3	6	4	4
3	4	12	6	8
3	5	30	12	20
4	3	12	8	6
5	3	30	20	12

What is astounding here is that for each pair (n, m) there is only one possibility for each of E, V, and F. Do such polyhedra exist? Here are the shapes and miniature dressmaker's patterns for them:

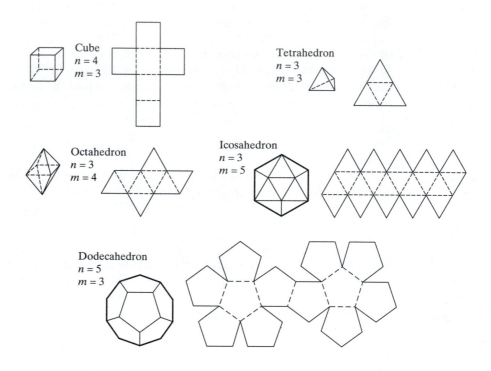

Except for the cube, the names of these polyhedra are Greek and refer to the number of faces each has: tetrahedron (four-faced), octahedron (eight-faced), icosahedron (20-faced), and dodecahedron (12-faced). [You can view these shapes in the appropriate The Geometry Files files.]

Problem 27   Large dressmaker's patterns for the cube and regular tetrahedron can be found in problems 2 and 5, respectively. The following are large dressmaker's patterns for the regular octahedron, regular dodecahedron, and regular icosahedron. Cut them out, fold, and assemble into the regular polyhedra.

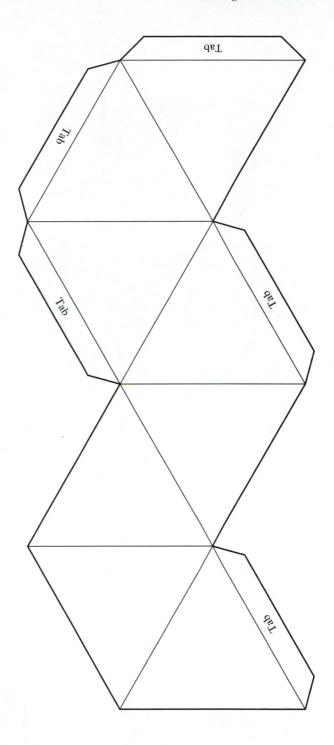

Dodecahedron

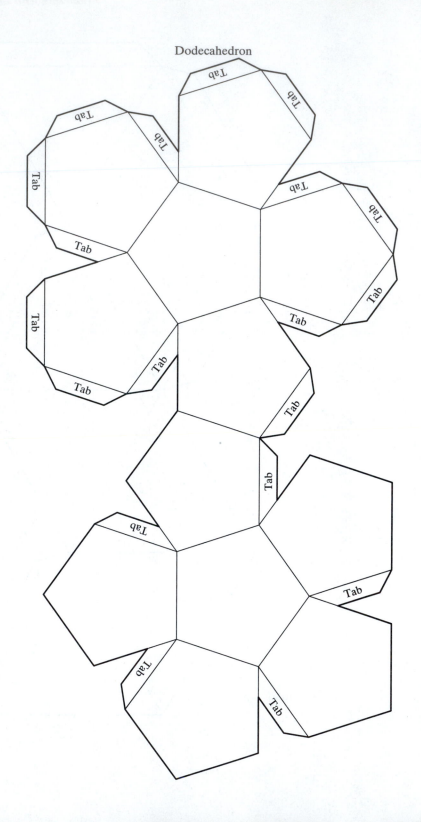

Icosahedron

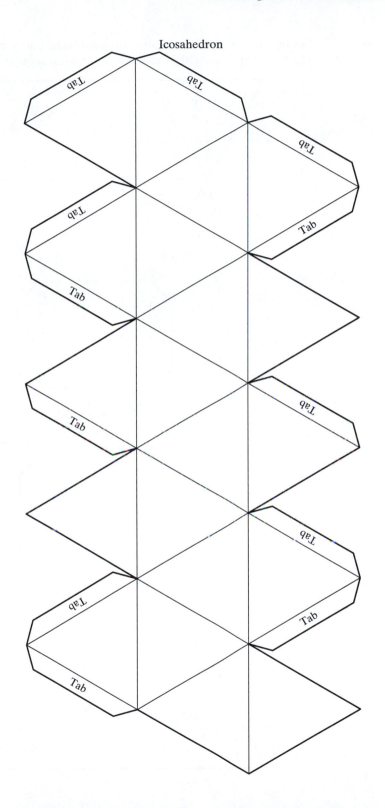

Problem 28   Cut out the patterns below, fold on solid lines, and assemble using glue or tape. The two shapes should fit together to make a regular tetrahedron.

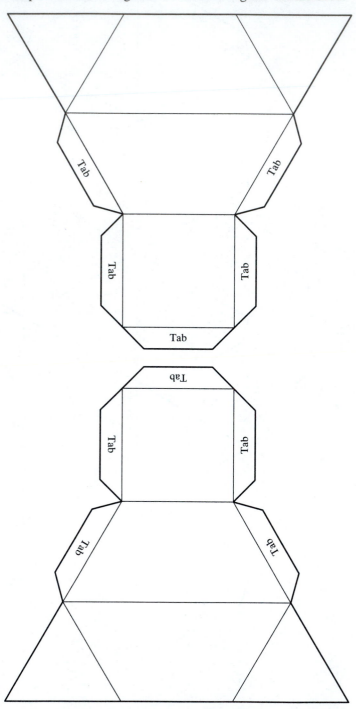

Problem 29   Intersect a cube with a plane. What polygons do you get? (Do you get any regular polygons?) Intersect a regular tetrahedron with a plane. What do you get? Intersect a regular octahedron with a plane. What do you get?

Problem 30   You can get a regular tetrahedron to **nest** inside a cube: the tetrahedron fits snugly in the cube without rattling around. All the edges of the tetrahedron are diagonals of the cube's faces. [To see this, construct two tetrahedrons inside a cube using The Geometry Files.]

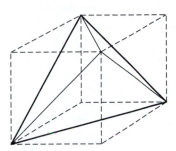

Suppose the cube has edge length equal to *s*. Find the volume of the regular tetrahedron that nests in the cube. Do it in two ways. For one way, use the formula for a triangular pyramid (another name for a tetrahedron). For the other, start with the cube and slice off the corners of the cube so that the regular tetrahedron remains. Volume of cube = sum of volumes of corners plus volume of tetrahedron. What is the ratio of the volume of the tetrahedron to the volume of the cube?

Problem 31   Show that you can get a cube to nest in a regular dodecahedron. In how many ways can this be done?

Problem 32   You can construct a regular octahedron inside a cube in the following manner. Join the centers of adjacent faces of the cube with a line segment. The union of these line segments forms the skeleton of edges of an octahedron. [To see this, construct an octahedron nesting in a cube with The Geometry Files.]

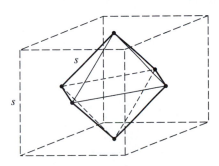

Suppose the regular octahedron has side *s*. Find its volume and do it in two ways. In one way, dissect the octahedron into two Egyptian pyramids. In the other, assume the octahedron sits inside the cube as in the picture above. Slice off pieces of the cube so that the regular octahedron remains. Volume of cube = sum of volumes of pieces plus volume of octahedron. What is the ratio of the volume of the cube to the volume of the octahedron?

Problem 33

To construct a truncated (regular) octahedron, start with a regular octahedron and remove caps at each vertex. Each cap is an Egyptian pyramid. Do this in such a way that the resulting faces of the new polyhedron are regular hexagons (what's left of the faces of the original octahedron) and squares (the bases of the Egyptian pyramids). [A good exercise would be to construct the truncated octahedron in The Geometry Files.]

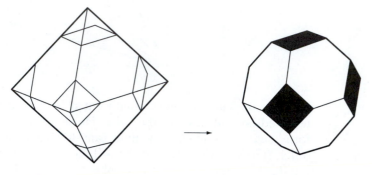

Truncated octahedron

All the edges of a truncated octahedron have the same length. (Why so?) What is the volume of a truncated octahedron whose common edge length is *s*?

✳ *The Soccer Ball*

A soccer ball is sort of a polyhedron. It's made of pieces of leather, several of which are congruent (curvy) regular pentagons and several of which are congruent (curvy) regular hexagons. The pieces are sewn together so that each pentagon is surrounded by hexagons and each hexagon is surrounded, alternately, by three

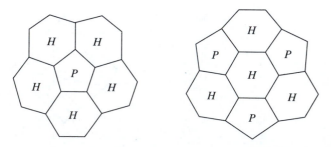

pentagons and three hexagons. Does this information alone determine the soccer ball uniquely? Could there be another polyhedron having these properties?

Suppose the information given is all we know about a soccer ball. Let's see if we can use Euler's formula and some clever counting (like we have used earlier) to predict additional information about the polyhedron.

As before, let V, E, and F denote the number of vertices, edges, and faces of the polyhedron. Let F_5 and F_6 denote the number of pentagonal and hexagonal faces, respectively, the polyhedron has. Thus, $F = F_5 + F_6$.

Now, each pentagon is surrounded by hexagons; five hexagons border on each pentagon. Thus, if we multiply the number of pentagons by 5, we will obtain a count of the number of hexagons. But the quantity $5F_5$ counts each hexagon more than once, in fact once for each pentagon it borders. Since each hexagon borders on exactly three pentagons, each hexagon will have been counted in $5F_5$ exactly three times. Consequently,

$$5F_5 = 3F_6$$

Also, at each vertex, two hexagons and one pentagon meet. Thus, there are exactly three edges attached to each vertex and

$$3V = 2E$$

Furthermore,

$$5F_5 + 6F_6 = 2E$$

(See the proof of Theorem 2 for a justification of both these equations.) Combining these equations, we obtain

$$3V = 2E = 5F_5 + 6F_6 = 9F_6$$

Now to plug all of this into Euler's formula:

$$12 = 6V - 6E + 6F = 18F_6 - 27F_6 + 6(F_5 + F_6) = -9F_6 + 6F_5 + 6F_6$$

$$= -3F_6 + 6F_5$$

$$= -5F_5 + 6F_5$$

$$= F_5$$

Thus, there are exactly 12 pentagonal faces. Working backwards through all the equations, we get that

$$F_5 = 12 \quad F_6 = 20, \quad E = 90, \quad V = 60, \quad F = 32$$

Problem 34   A **semiregular** polyhedron (sometimes called an **Archimedean solid**) has all its vertices congruent and all its faces regular polygons; but not all of the faces are congruent. Show that the soccer ball polyhedron described above is a semiregular polyhedron.

Problem 35   The faces of a certain convex polyhedron are all squares and regular hexagons. Each square is surrounded by hexagons, and each hexagon is surrounded alternately by three squares and three hexagons. Say as much as you can about the numbers of squares, hexagons, edges, and vertices of the polyhedron. Is the polyhedron semiregular? (See the definition of semiregular in problem 34.)

Problem 36   Several examples of polyhedra have been mentioned in this chapter, in the text, and in the problems. Which ones are semiregular? Can you think of any more? (See definition of semiregular in problem 34.)

A famous unsolved problem of geometry: Is it true that every convex polyhedron can be unfolded into a pattern in at least one way?

Problem 37   The equilateral triangle and the following parallelogram can both be folded into regular tetrahedra. Show this!

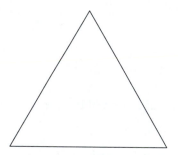

Problem 38   (A sequel to problem 37.) Can every parallelogram and every triangle be folded into (not necessarily regular) tetrahedra? If not, describe those parallelograms and triangles that can be folded into tetrahedra. Can a triangle or parallelogram be folded into a tetrahedron in more than one way?

Problem 39   (A sequel to problems 37 and 38.) Can any tetrahedron be obtained by folding a parallelogram or triangle? If not, describe those tetrahedra that can be folded from a parallelogram or triangle.

Problem 40   In problem 30, we saw that a regular tetrahedron nests nicely in a cube so that the edges of the tetrahedron are face diagonals of the cube. Actually, there are two such tetrahedra. (If you think about it, a cube has six faces, and, therefore, 12 face diagonals. A tetrahedron has six edges and so uses up half the face diagonals of the cube.) Make a model or use easy to follow graphics to show the two tetrahedra nested in a cube. Alternatively, do this on The Geometry Files.

Problem 41

(A sequel to problem 40.) The union of the two regular tetrahedra nested in a cube is a nonconvex polyhedron called the **stella-octangula.** What fraction of the volume of the cube does the stella-octangula occupy?

Problem 42

(A sequel to problems 40 and 41.) Make models of the stella-octangula and the cube it nests in. (Make the cube separately. Make it with a "lid" so you can place the stella-octangula in it to see it "nest.")

Problem 43

(A sequel to problems 40, 41, and 42.) Make models of the polyhedral spaces formed between the stella-octangula and the cube it nests in. Is there any way these polyhedra could be combined to make another stella-octangula?

Problem 44

We've seen that one way to create a new polyhedron out of an existing one is by truncation. (See problems 10 and 33). Another way to create a new polyhedron from an existing one is to choose a face of the existing polyhedron, then build a pyramid on that face as base. Of course, if the existing polyhedron is convex and you want the new polyhedron to be convex, too, then you have to make sure that the pyramid you attach is not too pointy. One famous polyhedron obtained in this way is called the **rhombic dodecahedron.** To each face of a cube, attach an Egyptian pyramid. The height of each pyramid must be half the height of the cube; the apex of each pyramid should lie directly above the center of the cube's face. Make a model of the rhombic dodecahedron. Look at what you get. Alternatively, you might want to construct this with The Geometry Files. Why is the model called a rhombic dodecahedron?

Problem 45

In Problem 32, we constructed a regular octahedron inside a cube by joining the centers of adjacent faces of the cube with a line segment. The union of these line segments forms the skeleton of edges of an octahedron. What happens if you start with an octahedron and do the same thing (that is, join the centers of each pair of adjacent faces of the octahedron with a line segment)? Then look at the polyhedron whose edge skeleton is formed by these line segments. Can you identify the polyhedron? Do the same thing with the other regular polyhedra: the regular tetrahedron, dodecahedron, and iscosahedron. What happens?

✳ *Full Circle*

Our major task in this chapter has been to explore the idea of polyhedra, the three-dimensional counterparts to polygons in the plane. A lot of effort was put into just getting acquainted with these objects. For example, you can draw a "model" of a polygon on a piece of paper. All you need to know about that polygon is basically right there in front of you. But a model of a polyhedron has to be constructed, and this takes time and patience. Polyhedra turn out to be more complex than polygons; we had to develop language with which to talk about them. Nevertheless, we were able to prove two startling theorems. One limited the possibilities for regular polyhedra in a surprising way. The other connected the number of vertices, edges, and faces of a polyhedron in Euler's formula. We used the analogy between polygon and polyhedron to formulate definitions and to pose the questions to which these theorems are answers. In a third surprising theorem ("there must be at least one face with five or few edges"), however, we encountered a technique—a counting argument—we had never encountered with polygons. This suggests that further study of polyhedra will bring about more new ideas and methods for investigation.

✳ *Notes*

After the pyramids of Egypt, the second most famous collection of polyhedra in ancient times is the set of regular solids. The Greek mathematician Thaetetus (c. 415–369 B.C.) seems to have been the first to give them a mathematical treatment. Plato (427–347 B.C.),

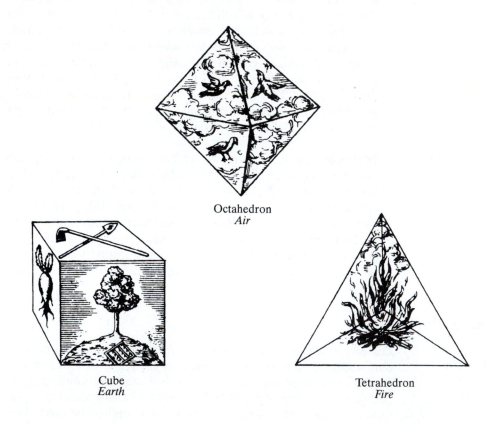

Octahedron
Air

Cube
Earth

Tetrahedron
Fire

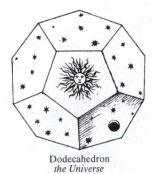

Dodecahedron
the Universe

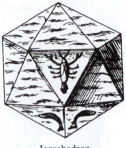

Icosahedron
Water

a friend of Thaetetus, incorporated the regular polyhedra into his cosmology. In his dialogue, *Timaeus,* he discusses the four "elements" of which everything is composed: earth, air, fire, and water. Earth particles have the form of cubes which stand solidly on their bases; air particles have the form of regular octahedra which are light and rotate freely when held by opposite vertices. Fire particles have the form of regular tetrahedra which have sharp corners; water particles have the form of regular icosahedra which are almost spherical and roll around like liquid.

By the time Euclid wrote his *Elements* (c. 300 B.C.) the Greeks had a full-blown theory of solid geometry. In Book XI of the *Elements,* Euclid discusses metric properties of polyhedra; in Book XIII, he showed how to construct the regular polyhedra and "proved" that there are only five of them. Heron (c. 75 A.D.?) was the first to refer to the regular polyhedra as the **Platonic solids.**

Pappus (320 A.D.) gives an account of a work of Archimedes (c. 287–212 B.C.), now lost, devoted to a study of the semiregular polyhedra, hence the alternative name, Archimedean Solids.

In the Renaissance, when the classical writings of Rome and Greece became available after being lost during the Dark Ages of Europe, the works of Plato and Euclid were studied by theologians, philosophers, artists, and scientists. Their study also brought about an interest in polyhedra.

Johannes Kepler (1571–1630) added to Plato's cosmology by giving the entire universe the shape of the dodecahedron, perhaps because its 12 faces correspond to the 12 signs of the zodiac. Each regular polyhedron now corresponded to a major facet of the world. Kepler went further than this and incorporated the regular polyhedra into Copernicus's system of planets moving in orbits about the sun, using them to explain the existence of exactly six planets (Mercury, Venus, Earth, Mars, Jupiter, Saturn) and the particular distances of these planets from the central sun. A young Kepler conceived of the idea that the five gaps between the six planets would correspond to the five regular solids, thus explaining two puzzles at once: why there are exactly five regular polyhedra and why there are exactly six planets. After trying many ways of arranging the regular polyhedra to fit this idea and the known data, he arrived at the scheme shown in the illustration on the next page. Saturn moves on the outer sphere. This contains a cube, in which is inscribed the sphere on which Jupiter moves. Jupiter's sphere contains a regular tetrahedron, in which is inscribed Mars's sphere. Similarly, Mars's sphere contains a dodecahedron, then the Earth's sphere, an icosahedron, Venus's sphere, an octahedron, and finally Mercury's sphere.

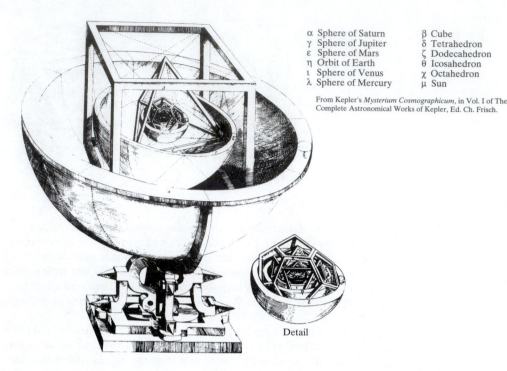

α Sphere of Saturn   β Cube
γ Sphere of Jupiter   δ Tetrahedron
ε Sphere of Mars   ζ Dodecahedron
η Orbit of Earth   θ Icosahedron
ι Sphere of Venus   χ Octahedron
λ Sphere of Mercury   μ Sun

From Kepler's *Mysterium Cosmographicum*, in Vol. I of The Complete Astronomical Works of Kepler, Ed. Ch. Frisch.

Detail

Kepler was sufficiently excited about his discovery that he proposed to his patron, the Duke of Würtemberg, that a gold model be constructed of the nested polyhedra and spheres to show the world his scheme for explaining the mysteries of the universe.

Some of Kepler's experience with polyhedra was not quasimystical. He was aware (through Pappus) of Archimedes's study of semiregular polyhedra and made a complete enumeration of these solids, giving an elaborate, case-by-case argument for the list's completeness.

During this period, polyhedra attracted the attention of many scholars, artisans, and artists—including Albrecht Dürer, who invented the notion of the dressmaker's pattern for a polyhedron, and Leonardo da Vinci, who illustrated a book on regular and semiregular polyhedra by Luca Pacioli.

Descartes (1596–1650) also studied polyhedra and proved a theorem one of whose quick consequences is Euler's formula. Leonhard Euler's discovery of his famous formula was first conveyed in a letter to Christian Goldbach in 1750. The first proof of the formula appears to be due to Adrian-Marie Legendre (1752–1833).

This does not end the story of polyhedra. Mathematicians study them still. Scientists continue to use them in describing the shapes of molecules, of crystals, and of organisms. Some of these connections are shown in the following pictures and diagrams.

Melencolia. An engraving by Albrecht Dürer (1471–1528).

A portrait of Fra Luca Pacioli and his student Guidobaldi, Duke of Urbino, by Jacopo de Barbari.

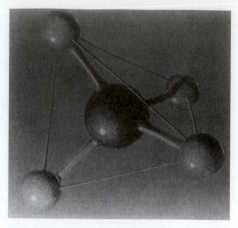

An artist's conception of a methane molecule from *The Architecture of Molecules* by Linus Pauling and Roger Hayward. W.H. Freeman, 1964.

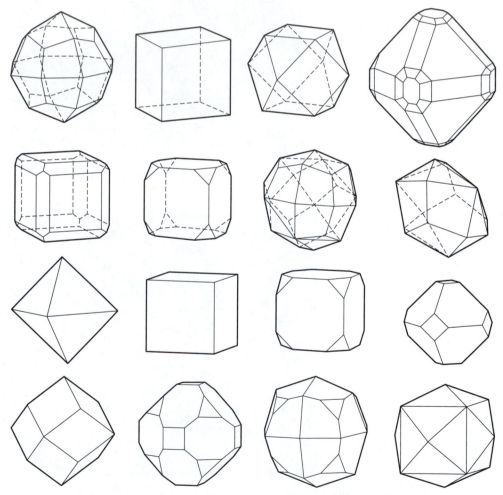

Drawings of crystals of gold, from *Atlas der Krystallformen* by Victor Goldschmidt.

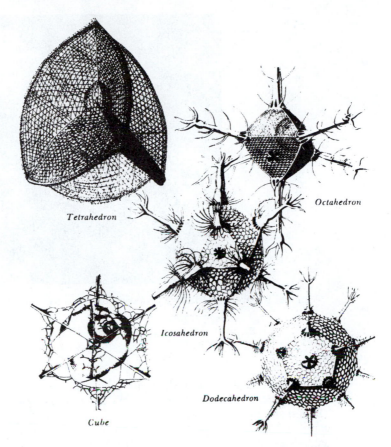

Good approximations of the regular polyhedra as skeletons of radiolaria, minute marine animals. From Ernst Haekle's *Report on the Scientific Results of the Voyage of H.M.S. Challenger,* Volume 18. London: 1887.

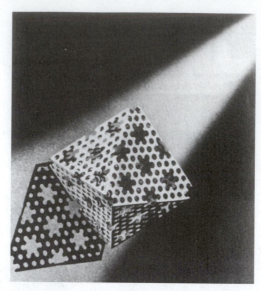

A model for protein structure proposed by Dorothy Wrinch in 1934. Smith College.

✳ **References**

Beard, Robert Stanley, *Patterns in space.* Palo Alto, CA: Creative Publications, 1973.

Beck, A., Bleicher, M. N., and Crowe, D. W., *Excursions into mathematics.* New York: Worth Publishers, 1970.

Cundy, H. M., and Rollett, A. P., *Mathematical models.* Oxford: Clarendon Press, 1961.

Edmondson, Amy C., *A Fuller explanation: The synergetic geometry of R. Buckminster Fuller.* Boston: Birkhauser, 1987.

Fejes Toth, L., *Regular figures.* New York: Macmillan, 1964.

Heath, Thomas Little, *The thirteen books of Euclid's Elements.* New York: Dover, 1956.

Heath, Thomas Little, *The works of Archimedes.* New York: Dover, 1953.

Holden, Alan, *Shapes, space and symmetry.* New York: Columbia University Press, 1971.

Kappraff, Jay, *Connections: The geometric bridge between art and science.* New York: McGraw-Hill, 1991.

Lines, L., *Solid geometry.* New York: Dover, 1965.

Loeb, Arthur L., *Space structures: Their harmony and counterpoint.* Reading, MA: Addison Wesley Pub. Co., 1976.

O'Daffer, Phares, and Stanley R. Clemens, *Geometry: an investigative approach.* Menlo Park, CA: Addison-Wesley, 1997.

Pearce, Peter and Susan, *Polyhedra primer.* New York: Van Nostrand Reinhold, 1978.

Pólya, George, *Induction and analogy in mathematics, Vol I in Mathematics and plausible reasoning.* Princeton, NJ: Princeton University Press, 1954.

Senechal, Marjorie, and Fleck, George M., *Shaping space: A polyhedral approach.* Boston: Birkhauser, 1988.

Steinhaus, Hugo, *Mathematical snapshots.* New York: Oxford University Press, 1950.

Wenninger, Magnus J., *Polyhedron models.* Cambridge: University Press, 1971.

4 Shortest Path Problems

Frequently, in our travels we are anxious to arrive at our destination by way of the shortest route. If the land is flat and the way is clear, then we assume that the shortest route is by a straight line path. But if it isn't flat or there are obstructions, then it may be another story. Also, it may be that our trip is not easily described as between two points. We are interested in all this because we don't want to expend more energy (or money) than we have to. Scientists are also interested in paths that use the least energy—for example, paths of light. Although this chapter's problems are built primarily around this theme, thinking about them and solving them will lead us to some surprising, related topics.

❋ The Speaker Wiring Problem

John was lying in bed one morning listening to his stereo. He imagined how nice it would be if he could place one of the speakers up near the ceiling in the corner opposite his turntable and amplifier, which were on a table in the corner just to his right where he could turn them on easily without getting out of bed. He was wondering how much wire he would need to connect the speaker to his amplifier. He knew that his room was shaped like a cube: a square floor 9 ft by 9 ft and a ceiling 9 ft high. He figured that the speaker-wire connection on the amplifier was 3 ft up from the floor in the corner and that the wire would attach to a point in back of the speaker exactly 3 ft down from the ceiling in the opposite corner. Help John figure out how long the wire should be if it takes the shortest route along the walls and floor of his room.

!
STOP! Before reading on, try this problem yourself.

Solution to the Speaker Wiring Problem.   John decides to draw a picture of his room, marking on it the two ends of the wire.

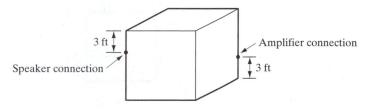

He thinks: "One way to do it would be to run the wire straight down the corner from the amplifier to the floor, across the floor diagonally, and up the opposite corner to the speaker.

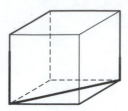

That would be, let's see, 3 ft from the amplifier to the ground plus $9\sqrt{2}$ ft across the diagonal of the floor (a little less than 13 ft) plus 6 ft up the corner to the speaker. In other words, roughly $3 + 13 + 6 = 22$ ft. Is that the shortest way? I know that if the wire were to run on a flat surface, then I could figure out the shortest way: I'd just draw a line between the two ends. Then I'd measure along the line from one end to the other to figure out the length. But my room is not flat! Is there a way I could make it flat? The ceiling, floor, and walls are flat themselves. The corners 'mess' things up. What if I were to 'cut' the room along the corners and flatten it out? I'd snip along the edges of the ceiling and bend it back, like this.

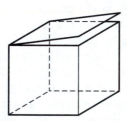

Then I'd snip along the corners from the ceiling to the floor, like this.

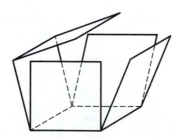

Then, I'd flatten it out, which would look like this.

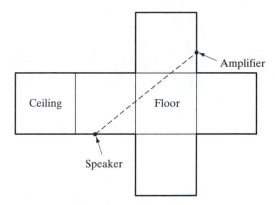

I could mark the two ends of the speaker connection and draw a straight line between the two points. Let me make a scale drawing of what I would get. From that I should be able to approximate pretty closely how long the wire should be. I'll get some graph paper and make the width of each little square on the graph paper represent a foot in the room. I won't have to include the ceiling.

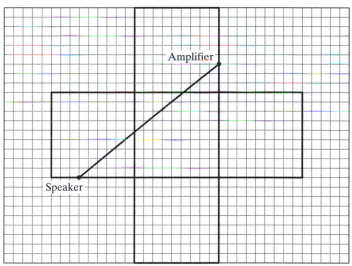

The side of each little square = 1 ft.

"On my scale drawing I've marked where the two ends of the wire will be and drawn the line between them. To measure the length of that line I'll take another piece of graph paper and use it to count the number of square-widths that will just fit along the line.

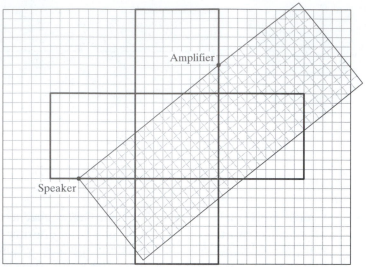

The side of each little square = 1 ft.

A little more than 19 squares. That represents a length of 19 ft in my room. That's shorter than the 22 ft I got before!''

Problem 1

Using the Pythagorean theorem, what is the exact length of the path shown in the preceding diagram? Is this really the shortest path? (In the diagram above you can also locate the speaker as being 3 ft down from the ceiling at the edge of the wall next to the one where it's marked now. This suggests another straight line path (on the flattened out cube) from speaker to amplifier. This may open up other, similar possibilities. . . Is it possible that unfolding the cube another way would result in other straight line paths?)

Problem 2

Below is a picture of a room. You want to install a light fixture right in the middle of the ceiling. You plan to run a wire from the fixture to the nearest electrical outlet, which is at the baseboard as shown. Find the shortest route for the wire to take and find the length of the route.

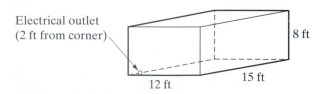

Electrical outlet
(2 ft from corner)

8 ft

12 ft   15 ft

Problem 3   A room is rectangular in shape, 18 ft long, 14 ft wide, and 10 ft high. (a) A mosquito flies along the shortest path from the southeast corner at the floor to the northwest corner at the ceiling. How far does it fly? (b) An ant walks from one of these corners to the other following a shortest path. How far does it walk?

Problem 4   You live in a geodesic dome, which, in this case, is an icosahedron with a five-triangle "cap" removed. (The "edge" of the cap is the floor of your house.) Each side of each triangular face has length 10 ft. Your CD and amp are located in the interior of a triangle at your dome's base, about 3 ft from the triangle's base (measured along the surface of the triangle). Your speaker is located on the triangle opposite, 3 ft down from **its** base. (The triangle opposite is inverted.) Find the shortest route for the wire from CD to speaker and its length.

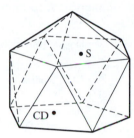

Problem 5   Take two points on the surface of a cylinder (ignore the circular top and bottom). Find the shortest path between them. Do the same thing with another pair of points. In general, investigate shortest paths between two points on the surface of a cylinder. Some pairs of points may have one kind of path, some another.

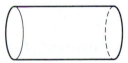

Problem 6   Same as problem 5 but for the surface of a cone.

Problem 7   Same as problem 5 but for the surface of a sphere.

Problem 8   (From *Calculus* by Deborah Hughes Hallett, et al.) Bueya spends her days as follows: She is at Washington University in the morning, at her job in East St. Louis in the afternoon, and late in the evening she has a beer in her favorite bar. She has lunch and dinner at home. Where on the road shown below should she look for an apartment to minimize her daily traveling distance?

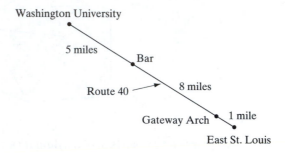

Her colleague Marie-Josée goes to a gym near the Gateway Arch before breakfast (which she eats at home), and she spends the rest of the day the same way as Bueya. Where on the road should **she** look for an apartment?

✳ *The Problem of the Milkmaid, the River, and the Cow*

Once upon a time there was a milkmaid who lived in a house on the banks of a river. On the same side of the river, but farther down, was her cow, tied to a tree. The milkmaid is about to milk her cow but before doing so she must first wash the milk bucket in the river. She knows that there are many paths from where she is to the river and then many paths from her washing spot to the cow. She finds this knowledge very interesting but wonders whether, among all these paths, there is a shortest one, and, if yes, what it is. Help her out.

!   ***STOP! Try this problem before reading on. Geometer's Sketchpad would be a good help in getting started with this problem.***

A Solution to the Milkmaid, River, and Cow Problem. She thinks: "Let me draw a picture.

"Then let me mark a point on the river where I might wash the bucket. Is it important where that point is? I know that the shortest distance between two points is a straight line so the path from my house to the washing point should be a straight line and the path from the washing point to the cow should be a straight line. The total length of my trip would be the sum of the two lengths.

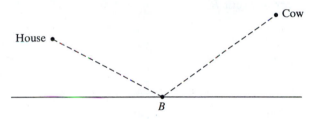

"The point on the river must be just a little bit important because if the point were at *A* then the total trip would be shorter than if the point were at *B*.

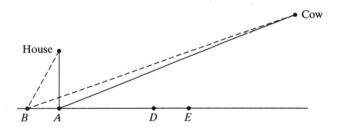

"But for points *D* and *E*, which total trip is smaller? Maybe the two trips have the same length! It's hard to tell. It would be so nice if the whole trip were a single straight line drawn between two points, and not a 'broken' line as it is now. The speaker problem was turned into one of connecting two points with a straight line. Could I do the same thing here?

"Hmm. If my house were on the other side of the river, then it would be easy. Just join house and cow with a straight line.

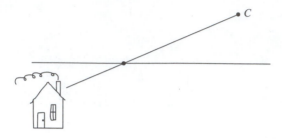

"Little bit of inconvenience: I'd have to walk through the river—or on top of it. . . . Wait! Let me make a copy, on the other side of the river, of my situation on this side. My house on this side is at *H*. I'll put a house directly opposite on the other side at *H'*. I'll make the distance of *H'* to the river the same as the distance of *H* to the river. Connect *H'* and the cow at *C* with a straight line intersecting the river at point *R*.

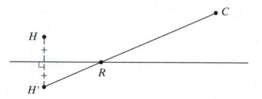

"If I join *H* to *R* on this side, I get a copy of the path from *H'* to *R*. Paths *HR* and *H'R* have the same length! (Side-angle-side, congruent triangles, corresponding sides—right?) That gives me a point on the river and a path from *H* to *R* to *C*. Is this the best I can do?

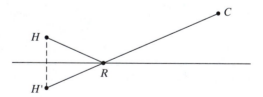

"Suppose I washed the bucket at another point *P* on the river. Then my trip would be *HP* plus *PC*. How does this compare with the other trip, *HR* plus *RC*? If I lived on the other side of the river and washed my bucket at *P*, my trip would be *H'P* plus *PC*—the same length as *HP* plus *PC*. (Why is this?) Then, since the shortest distance between two points is a straight line and the trip from *H'* to *R* to *C* is a straight line, the trip from H' to P to C is longer than the trip from H' to R to C. Consequently, since the latter trip has the same length as the trip from H to R to C, the trip from H to P to C is longer than the trip from H to R to C. Yes! My best trip is from H to R to C."

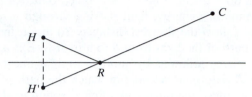

Problem 1 for Right Now!   Let X and Y be points up river and down river, respectively, from R, the point in the solution to the Milkmaid, the River, and the Cow Problem. Show that R has the property that angles HRX and CRY are congruent. Show also that R is the only point along the river-line having this property.

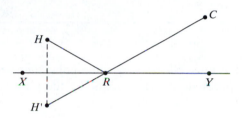

Problem 9   Two cities, Athens and Sparta, are situated on a plain near the River Nod as shown in the scale drawing map below. They plan to build a joint water treatment plant on the river. Pipes (in straight lines) from the plant to both of the cities will have to be laid. The cities want to locate the plant in the position along the river that makes the total length of the piping the smallest. On the map, indicate where the plant should be built and sketch in the pipelines. Label your sketch so that a civil engineer would know how to create an accurate scale drawing of the plant location and the pipelines.

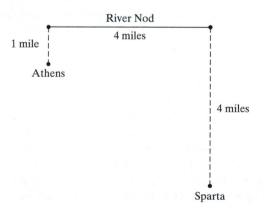

✳ *Paths of Light; Mirrors*

Suppose that in front of a flat mirror are an object (at point S) and your eye (at point E). Light from the object is reflected off the mirror and hits your eye. What path does the light take from object to mirror to eye? Fermat's principle says that, to minimize energy, light passing through a homogeneous medium (such as air) must follow the shortest distance. So in air, light must follow a straight line. Thus the path of the light from S to mirror to E is a straight line from S to a point P on the mirror followed by another straight line from P to E. The three points S, P, and E determine a plane in which the path of the light lies. This plane intersects the plane of the mirror in a line L giving us the following picture.

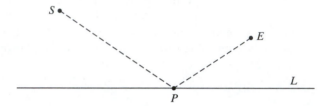

Point P must be the very same point that would solve the Milkmaid, River, and Cow problem, that is, P must be the same as that point Q on line L for which $SQ + QE$ is the smallest.

Problem 2 for Right Now!

Show that the plane of S, P, and E—the plane that contains the path of the light—must be perpendicular to the plane of the mirror.

As a consequence of the solution to the Milkmaid (et al.) Problem and of the results of Problem 1 for Right Now!, we have the following picture of the light path.

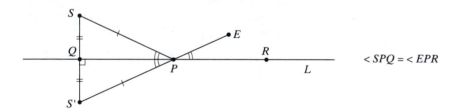

Assume the plane of the page is the plane of the light path. Then, according to Problem 2 for Right Now!, the mirror is perpendicular to the page and passes through line L. Any object we see can be considered as a light source. So the path of light from source to mirror to eye—whether it be from a pencil or a flashlight—is the same as in the diagram above.

What do we see when we look along the path PE toward the mirror? Our eyes are at E. Our eyes assume light is coming in a straight line path directly from its source, as it normally does. So, our eyes assume that light from S reflected from the mirror at P is coming from S', the "virtual" image of S. We see S, not just where it is, but also "behind" the mirror at S'. The point S' is the **mirror image** of the point S.

Problem 3 for
Right Now!

A flat mirror has another interesting property that we take for granted: No matter where our eye is relative to S or to the mirror (as long as it is above the reflecting side of the mirror), the point S' always appears to be in the same place. Why is this?

We can use this observation and the geometry of the diagram above to explain some of our experiences with mirrors. For example, place a mirror on a table and perpendicular to its surface. In front of the mirror, place a design on a piece of paper. As you look into the mirror, you see the mirror image of the design "in" the mirror (or "behind" it). The mirror image also "sits on the table" but the mirror image appears to be the result of having turned the original design over (upside down) and then viewing the original design "through" the paper.

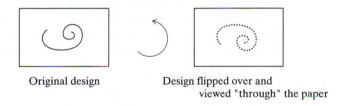

Original design

Design flipped over and
viewed "through" the paper

Here is what we imagine the scene would be if viewed looking down at the table from above.

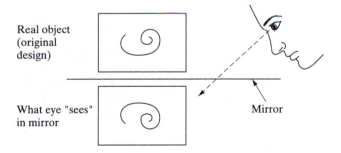

Real object
(original
design)

What eye "sees"
in mirror

Mirror

This is a picture of the original design and its mirror image joined together to make a new design. The same effect can be approximated by looking down on the table but slightly in front of the mirror.

Problem 10

Suppose you were to place the following design on a table with its edge snug up against the front of the mirror (as in the setup above with mirror perpendicular to table). Draw a careful picture of the total design you would get consisting of the original and its mirror image.

Problem 11   When you stand directly in front of a mirror and look, you see a mirror image of yourself. The image in the mirror has its right and left interchanged. If you raise your right hand, the mirror person raises its left. Why does the mirror interchange right and left? Why doesn't your mirror image have top and bottom reversed instead? Imagine a neon sign on a glass storefront, with a mirror behind it. Does the image of the sign read left-to-right or right-to-left? How is an image reversed in a camera? What about an image in your retina?

Problem 12   (a) Place two mirrors at right angles. A light ray is directed at one of the mirrors. What can you say about the subsequent path of light?

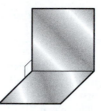

(b) Place three mirrors so that any pair are perpendicular—such as when three mirrors are placed in the corner of a rectangular room. Direct a ray of light at one of the mirrors. What can you say about the subsequent path of light? Three perpendicular mirrors are used in the red warning lights on bicycles and cars and also in radio communication. And, I've been told, in a gizmo that bounces a laser beam from the earth off the moon. Why is this?

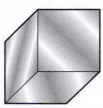

Problem 13   Place two mirrors at right angles, one flat on the table and the other perpendicular to the table. Look at yourself in the pair of mirrors, roughly at the "crack" where the two mirrors join, as in the diagram below. What do you see? Explain.

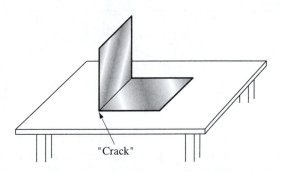

"Crack"

Problem 14   Take three mirrors that are mutually perpendicular as in problem 12 (b). Look at yourself in the mirrors, roughly at the point where the three mirrors meet. What do you see? Explain.

✳ *The Cowboy Problem*

In the picture below, a cowboy is at point *C* on his horse. Before moving to his campsite (point *S*), he first wants to take his horse to get some grass at the edge of the field, then go to the river for a drink. What is the shortest path from *C*, to the edge of the field, to the river and then to *S*?

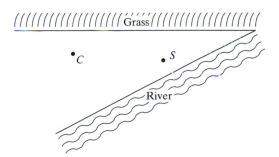

!   ***STOP! Try this before reading on. Use Geometer's Sketchpad to help you get started.***

   Solution to the Cowboy Problem   The critical points in a solution are *P* and *Q*, places where the cowboy stops at the edge of the field and at the river, respectively. To figure out where these points are, we try a tactic that worked with the milkmaid problem: we consider a "virtual" cowboy at *C'* (= reflection of *C* in a "mirror" placed along the edge of the field) and a "virtual" campsite at *S'* (= reflection of *S* in a "mirror" placed at the river). Then we draw the line connecting *C'* and *S'*.

This line intersects the edge of the field at P and the river at Q.

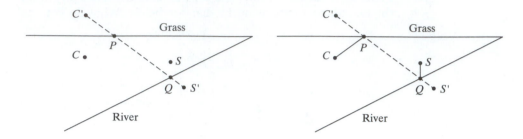

We claim that the shortest trip is C to P to Q to S. To prove this, we first note that the length of this trip is equal to the distance from C' to S'. Now suppose the cowboy takes another trip stopping at P' and Q' on the edge of the field and the river, respectively. Reflect CP' in the mirror at the edge of the field and get $C'P'$. Reflect $Q'S$ in the mirror along the river to get $Q'S'$. The length of the trip passing through P' and Q' is

$$CP' + P'Q' + Q'\text{S}$$

which is equal to

$$C'P' + P'Q' + Q'S'$$

which in turn is greater than $C'S'$ because the path $C'P'Q'S'$ is not a straight line path from C' to S'!

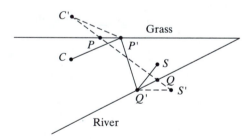

An optical interpretation of the solution to the cowboy problem is the following. If the edge of the field and the river are replaced by **real** mirrors, then a ray of light from C to P hits S after two reflections. Similarly, if the two mirrors are replaced by two sides of a billiard table, then a billiard ball struck at C toward P will bounce off the two sides and hit S.

Problem 15   Suppose the cowboy takes his horse to the river first, then to the edge of the field, and finally to S. What should be the path? Is the path longer or shorter than the path gotten by going to the field first? Can you say anything about when the two paths are the same size? Geometer's Sketchpad might be useful here.

Problem 16   Imagine the cowboy again on his horse at C. This time he wants to go the edge of the orchard first to pick an apple for his horse, then to the edge of the field to let his horse eat some grass, then to the river to give his horse a drink, and finally to his campsite S. What is his shortest path? Why (natch)?

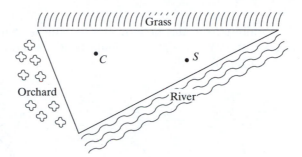

❋ Ellipses

Amazingly enough, some of the things we have been discussing have implications for an ellipse! Before seeing what these are, let's take time to recall just what an ellipse is. To determine an ellipse you need two points A and B in a plane and a fixed number α bigger than the distance AB. The set of points X such that $XA + XB$ equals the fixed number α is an ellipse. The points A and B are the ellipse's *foci* (singular: *focus*).

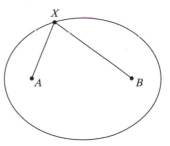

Problem 17   *An apparatus for drawing an ellipse.* Pound nails into a plane at points A and B. Take a piece of string whose length α is greater than AB; tie its ends to A and B. Keeping the string stretched tight with your pencil, draw a curve around the two nails as in the diagram below. The curve will be an ellipse with foci A and B, satisfying the equation $XA + XB = \alpha$ for every point X on the curve.

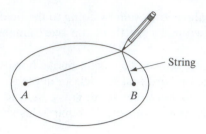

It follows from this definition of an ellipse that if Y is a point in the ellipse's interior, then $YA + YB < \alpha$. Similarly, if Y is a point in the ellipse's exterior, then $YA + YB > \alpha$.

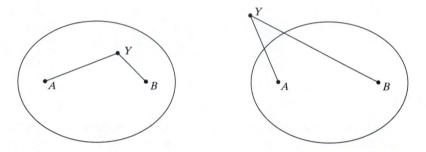

Choose a point R on the ellipse and let HJ be the tangent line to the ellipse at R. Consider the milkmaid problem with one focus A playing the role of the milkmaid, the other focus B playing the role of the cow, and the line HJ playing the role of the river. What is the point Q on HJ that minimizes $AQ + BQ$?

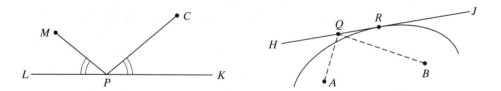

We claim that $Q = R$. If this were not the case, then—since the tangent line HJ intersects the ellipse in the one point R—the point Q would be on the outside of the ellipse. Therefore, $AQ + BQ > \alpha = AR + BR$, contradicting the minimality of $AQ + BQ$ for points on the line HJ. Thus $Q = R$. Furthermore, from Problem 1 For Right Now!, $\angle ARH = \angle BRJ$. This gives us the following result.

Theorem 1. Given an ellipse with foci A and B, point R on the ellipse, and tangent line HJ to the ellipse at R. Then $\angle ARH = \angle BRJ$.

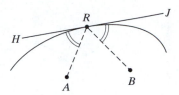

Problem 18   Take a point P on an ellipse having foci A and B. Use the theorem to construct (with ruler and compass) the tangent line to the ellipse at P.

This theorem has an interesting consequence. Suppose you bend a mirror into the shape of an ellipse. The mirror will be perpendicular to the plane of the ellipse. Then place a light source at A, one of the foci. When light from this source travels to a point R on the ellipse, it will be reflected from the mirrored ellipse just as it would be reflected from a "straight" mirror coinciding with the tangent line to the ellipse at R. (Think of the tangent line at a point as the best straight line approximation to the curve at that point.) Since $\angle ARH = \angle BRJ$, the light will be reflected back from R to the other focus B.

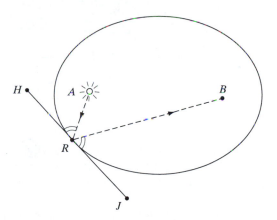

If you revolve an ellipse about its major axis (the line containing its two foci), then you obtain a surface called an *ellipsoid*. Make a mirror out of this surface by covering the inner, concave side with polished metal. All light originating at one focus of the ellipsoid will be reflected to the other focus. A dentist's lamp uses a piece of this mirrored ellipsoidal surface to focus light at one spot in the patient's mouth.

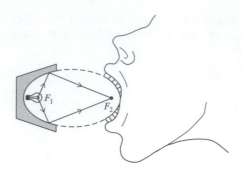

Suppose you have an object that needs to be illuminated from all sides. Build an elliptical mirror and place the object at one focus and a lamp at the other. Then turn on the switch.

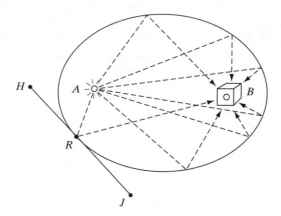

Sound reflects off a surface just as does light. The Capitol building in Washington, D.C., contains an elliptically shaped "whispering gallery," designed so that a message whispered at one focus can be heard at the other, but nowhere else in the room.

The reflection property of the ellipse has made possible the development of a treatment for kidney stones known as *Extracorporeal Shockwave Lithotripsy*. The device that does this is called a *lithotripter* (literally, stone crusher). The following description of this device appears in the article by Charles G. Moore listed at the end of the chapter: In this treatment,

> . . . the patient is first lowered into a special tub of warm water. At the bottom of the tub is a reflector cup in the shape of one end of an ellipsoid as described above. An electrode is fixed with its gap at one focus of the ellipse.

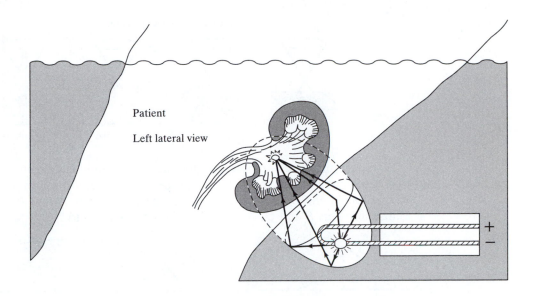

The patient is then positioned, with the aid of a fluoroscope (a machine which enables the imaging of internal body organs), so that the kidney stone is precisely at the other focus. When a spark is jumped through the gap of the electrode, the water at the gap is vaporized, causing a high-energy shock wave to radiate from the gap toward the reflector. The reflection property of the ellipse causes the shock waves to pass through the water, through the resilient tissue of the body, and to concentrate on the kidney stone at the other focus. The tissue in the neighborhood of the stone is not damaged. Since the acoustical properties of water and body tissues are almost identical, the shock wave passes unimpeded between the two foci of the ellipse.

It sometimes takes 2,000 shocks delivered over a half-hour to break up a stone into sand-sized particles that can be passed through the urinary system. The patient can expect to recover from the lithotripsy treatment in a matter of days as compared with a recuperation period of up to six weeks if the stone were to be removed surgically.

✳ *Parabolas*

You can think of a parabola as a limiting case of an ellipse. Take a cone and intersect it with a plane to make a circular cross section. (You imagine the cone to extend infinitely downwards.) Slowly tilt the plane. At first the cross sections will be ellipses. Then, suddenly, just when the plane becomes parallel to one of the slant sides, the cross section will be a parabola. If you keep tilting the plane, the cross section will become a hyperbola. (See the notes at the end of this chapter for a discussion of the connection between the Greek "conic sections" with the modern analytic geometry definitions of ellipse, parabola, and hyperbola.)

Cross sections of a cone

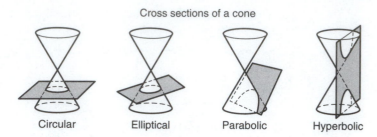

Circular   Elliptical   Parabolic   Hyperbolic

You can think of a parabola as a limiting case of an ellipse in another way, one that takes into account the reflective property of the ellipse. Start with an ellipse having foci A and B. Let the major axis (the line joining A and B) intersect the ellipse in points P and Q. Assume P is the point closest to A.

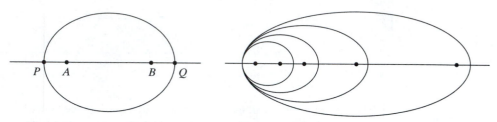

Allow the ellipse to change by keeping A and P fixed but moving the other focus B further and further away from A along the major axis. The point Q will move as well. As B moves, you should see an ellipse that is getting longer and longer. Imagine (as we did before) a flexible mirror perpendicular to the plane molded to the curve of the ellipse. Place a light source at A. The light will be reflected off the "mirrored" ellipse to the other focus at B. Then as B moves away from A, the light from A will be reflected off the ellipse in a direction more nearly parallel to the major axis. In the limit, the direction **is** parallel.

In the limit, the ellipse also becomes a parabola with vertex P and focus A. This seems plausible, but not entirely obvious. Let's show that this is so analytically in the following argument.

We will set the ellipse in the x-y plane and display its familiar equation. Place the ellipse so that its major axis is on the horizontal x-axis and its leftmost point is at the origin. Assume that its foci A and B have coordinates $(a, 0)$ and $(a + b, 0)$, respectively.

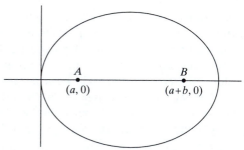

The equation for an ellipse with center $(c, 0)$, major axis of length $2M$ and minor axis of length $2m$ is

$$(x - c)^2/M^2 + y^2/m^2 = 1$$

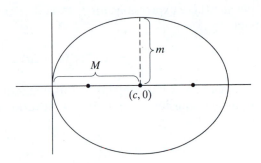

Let's figure out what c, M and m are in terms of a and b. First of all, c and M must be equal. Second, the center of the ellipse must be half-way between the two foci. Therefore,

$$c = M = (b + 2a)/2$$

To determine m from a and b, consider the following diagram.

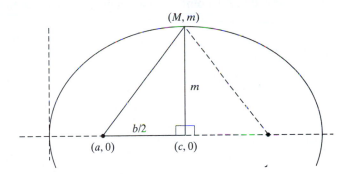

Note that m is a leg of a right triangle with vertices (M, m), $(c, 0)$, and $(a, 0)$. We know the leg joining $(a, 0)$ with $(c, 0)$ has length $b/2$. What about the length of the hypothenuse, the side joining $(a, 0)$ with (M, m)? We use the property of the ellipse: The sum of the distances from the point (M, m) to the two foci must be a fixed number. Moreover, the distance from $(a, 0)$ to (M, m) must be half this. But what is the fixed number? Is there a point on the ellipse, the sum of whose distances to the foci we know? Yes! the point $(0, 0)$ is on the ellipse. Its distance to $(a, 0)$ is a (natch), and its distance to $(a + b, 0)$ is $a + b$ (natch). The sum is $2a + b$. Thus the distance from $(a, 0)$ to (M, m) is $(2a + b)/2$. Using the Pythagorean theorem and a little algebra, we get that

$$m^2 = [(2a + b)/2]^2 - (b/2)^2 = a^2 + ab$$

Replacing these expressions for c, M, and m in the original equation for the ellipse, we get

$$(x - (b + 2a)/2)^2/[(b + 2a)/2]^2 + y^2/(a^2 + ab) = 1$$

We want to see what happens to this equation as b gets large. Before doing this, let's pause to consider the parabola with vertex at the origin and focus at $A = (a, 0)$.

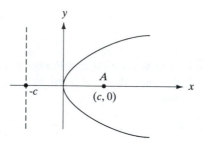

To find the equation for this parabola, recall that a point (x, y) is on this curve if its distance to the point $(a, 0)$ is equal to its distance to its **directrix**, the line $x = -a$:

$$x + a = \sqrt{(x - a)^2 + y^2}$$

or

$$(x + a)^2 = (x - a)^2 + y^2$$

or

$$x^2 + 2ax + a^2 = x^2 - 2ax + a^2 + y^2$$

or

$$y^2 = 4ax$$

We claim that this is the equation we will get when we let b go to ∞ in the equation for the ellipse. Let's see. The equation for the ellipse is

$$(x - (b + 2a)/2)^2/[(b + 2a)/2]^2 + y^2/(a^2 + ab) = 1$$

or

$$y^2 = (a^2 + ab)\{1 - (x - (b + 2a)/2)^2/[(b + 2a)/2]^2\}$$
$$= 4(a^2 + ab)\{-x^2 + (2a + b)x\}/(2a + b)^2$$
$$= \{-4(a^2 + ab)/(2a + b)^2\}x^2 + \{4(a^2 + ab)/(2a + b)\}x$$

It's easy to see that the limit of the right-hand side of this equation as $b \to \infty$ is $4ax$. Thus, if you take the ellipse with foci A and B and let the focus $B \to \infty$, you get a parabola with equation $y^2 = 4ax$.

The latter is the equation for a parabola with focus $(a, 0)$ and vertex $(0, 0)$. Interpreting this in terms of reflections, we see that a light source placed at the focus of a parabolic mirror reflects off the mirrored sides of the parabola outward in a direction parallel to the horizontal axis which is also the axis of the parabola. From another point of view, a ray of light coming into the mirrored parabola from the outside, in a direction parallel to the axis of the parabola, will reflect off a side of the parabola and be directed to the focus of the parabola. This also proves the following theorem:

Theorem 2. Given a parabola with focus A, point P on the parabola, tangent line BPC to the parabola at P, and line PD parallel to the parabola's axis. Then $\angle BPA = \angle DPC$.

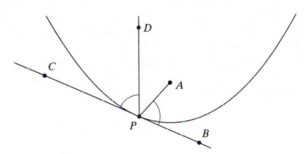

Problem 19   Given a point P on a parabola with focus F and directrix D, describe a method for using ruler and compass to construct the tangent line to the parabola at P.

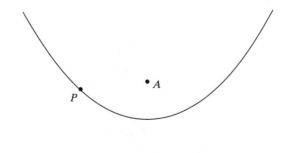

Problem 20

(From Jennings, 1994, *Modern Geometry with Applications.*) The following describes a method for drawing parabolas. Make a "T-square" by joining two boards or rods at right angles to form the letter "T." Cut a string the same length as the trunk of the "T."

On a piece of paper draw the directrix D and focus F of the parabola to be constructed. Attach one end of the string to the bottom of the trunk of the "T" and the other at F. Place the crosspiece of the "T" along the directrix. By sliding the crosspiece along the directrix, pulling the string tight with your pencil and keeping the pencil in contact with the trunk, trace out a curve. The curve should be part of a parabola with focus F and directrix D. Why is this?

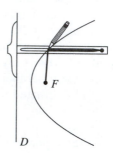

Some solar water heaters are based on the reflection property of the parabola. A flat piece of reflective material is bent into the shape of a parabolic "trough," i.e., one having parabolic cross section. A pipe through which water flows is placed along the line determined by the foci of the cross-sectional parabolas. When the axis of the parabola is pointed at the sun, the rays of the sun will be basically parallel to the axis, will reflect off the reflective sides of the trough, pass to the pipe located at the focus, and heat the water.

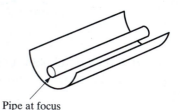

Pipe at focus

Other devices based on the reflective properties of the parabola involve the surface created by revolving a parabola about its axis.

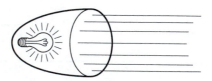

Solar ovens and lighthouse and high-beam automobile lights are designed in this way. The interior is coated with a reflective material. A solar oven works basically like the solar water heater above. With the axis of the parabola pointed at the sun, the sun's rays hit the reflected surface, pass to the focus, and cook the food placed there. However, for the lighthouse and automobile headlights the light source is placed at the focus. Light from the focus is reflected outward off the surface in a direction parallel to the axis of the parabola. This reflected light forms a cylindrical beam of light whose diameter is equal to that of the "rim" of the parabolical "bowl." A satellite receiver is also in the shape of a parabola of revolution and acts more like the solar cooker. Radio waves enter the reflector from the outside, reflect off a smooth surface, and are gathered at the focus.

✳ *The Three Cities Optical Network Problem*

You have been hired by a consortium of three cities to help them design a communications network. The three cities—Alphaville, Brookton, and Catalina—are thinking about how they might link themselves together using optical fiber. A linkage between two locations is constructed by laying an underground cable joining the two points. The total cost of these linkages is the cost of the cable itself and the cost of digging the trenches in which to lay the cable. Fortunately, the three cities lie on a plain so that the terrain for laying the cable is fairly uniform throughout the region. Thus, they figure that the cost of the project would be the cost of a unit length of cable and its installation times the total length of the network. But they figure that there are many ways to link up the three cities. Your job is to sort out all these ways and find a network connecting the three cities whose total length is smaller than other networks.

○
Alphaville

○
Brookton

○
Catalina

!   *STOP! Try this before reading on.*

Getting started with the Three Cities Problem. You try the obvious: Connect all the cities together with straight lines to form a triangle.

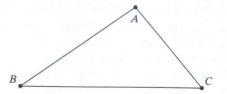

(You abbreviate the three cities by the letters *A*, *B*, and *C*.)

You realize that all three sides of the triangle are not necessary: For example, *A* can be connected to *C* without a **direct** connection:

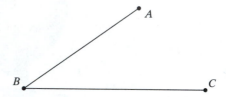

You think: Which side of the triangle should I leave out in order that the sum of the lengths of the remaining sides be the shortest? Clearly, you want to leave out the longest one.

Problem 21

To figure out which side of the triangle is the longest all you have to do is measure the three sides. Thinking about future jobs for which the configuration of cities might differ from this one, you ask if there is any other way to tell which side of a triangle is longest. Is there?

"Good. Making progress," you tell yourself. But then: "Can I do better? Is there some other way to hook up the cities so that the total length of the network is even smaller? Right now city *B* is kind of a hub; *A* and *C* are both connected to *B* directly. What if I moved the hub just outside *B*, in the general direction of *A* and *C*? I'd need an additional short link-up from the hub to *B*, but it looks like the connections from *A* and *C* to the hub would both be shortened. Would they be shortened enough to make up for the added linkage from hub to *B*?

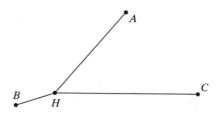

"Hmm. I could keep going. Move the hub H even further from B. But I don't want to move it **too** far. Interesting. Looks as if I'm looking at a network with **four** nodes: A, B, C, and H. What I want to know now is where H should be located in the triangle so that $HA + HB + HC$ is the smallest."

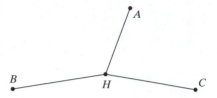

You think: What do I know about special points in a triangle? You reflect on your previous experience with geometry. Out of the deep past comes a voice: "The angle bisectors of a triangle meet in a point." Other voices shout out messages about competing points: "The medians of a triangle meet in a point!" "The altitudes of a triangle meet in a point!" "The perpendicular bisectors meet in a point!" "Look at the circumcenter!" "No, try the in-center!"

Computer Experiment

Use Geometer's Sketchpad to search for the best hub H in triangle ABC. Use what you know so far. Experiment with different types of triangles: equilateral, isosceles (but not equilateral), scalene triangle, 3-4-5 right triangle, triangle with a **big** angle (i.e., one whose measure is more than 125°).

Classroom Experiment

The idea is to take several types of triangles, construct in each one the special points mentioned above, measure the distances from A, B, and C to each point, and finally compare to see (for each triangle) which special point would make the best optical fiber hub. One way to do this is to break the class up into groups of three or four students each. Give each group a large sheet of newsprint paper (better yet, a large piece of butcher paper), rulers, compasses, and protractors. Have the group draw (carefully!) a large triangle on its piece of paper, construct the special points, make the measurements, and compare. Assign each group a different type of triangle (equilateral, isosceles nonequilateral, scalene nonisosceles, 3-4-5 right triangle, triangle with angle bigger than 125° . . .). Before breaking into groups, maybe the whole class could do the equilateral triangle. At the end, compare results and observations.

Another Approach to the Three Cities Problem. Let's try another approach to the problem. Let's assume we have a point H such that $AH + BH + CH$ is smallest and see if we can say something about H. There are two possibilities: (1) H is one of the cities and (2) H is not one of the cities. In case (1), we know which of the cities it must be: It's the city opposite the longest side.

Let's suppose that it's case (2): the point H where $AH + BH + CH$ is smallest is **neither** A nor B nor C. What can we say about H? Consider two of the cities:

B and *C*. The sum *BH* + *CH* suggests an ellipse: the set of all points *P* such that *BH* + *CH* = *BP* + *CP*. All such points *P* form an ellipse through *H* having foci *B* and *C*.

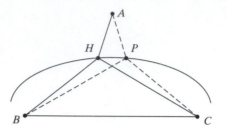

At what point *R* on the ellipse is *AR* a minimum? Yes. At the point where *AR* is perpendicular to the ellipse, or, equivalently, where *AR* is perpendicular to the tangent line to the ellipse at *R*. Thus, *P* and *R* must be the same point; no other point has this minimum property.

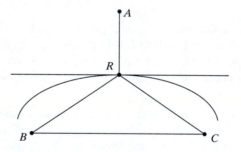

Moreover, from the reflection property of the ellipse, we have

$$\angle SHB = \angle THC$$

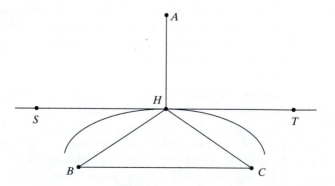

Hence,

$$\angle AHB = \angle AHC$$

Similarly, starting the argument with A and B instead of B and C, we can show

$$\angle CHB = \angle AHC$$

Putting all of this together, we conclude that all three angles $\angle AHC$, $\angle AHB$, and $\angle CHB$ are all equal in size. Since the three angles must sum to 360°, we must have that each one measures 120°! (Use protractors to check this requirement against the magic points of the big triangles you constructed in the preceding Classroom Experiment.)

We haven't solved the problem yet. We know which city it is if the solution point is one of the cities. And we know something about the geometry of the point (and the roads) if the solution point is not one of the cities. Maybe we can decide later just when each of the two cases is applicable.

Problem 22 (a) Show that the solution point can never be exterior to the triangle ABC. (b) Show that, if triangle ABC has an angle bigger than or equal to 120°, the solution point must be one of the cities.

✳ *How to Construct the Special Point for the Three Cities Problem?*

Given points A, B, and C, how can we construct the point H such that the three angles $\angle AHB$, $\angle BHC$, and $\angle CHA$ are all equal to 120°? Let's work backwards again and assume we have found the point. Perhaps the geometry of the situation will tell us how we could have gotten there. The setting looks like this:

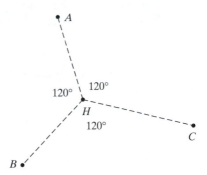

You can draw a circle through A, B, and H (there is only one such circle). And you can also draw a circle through A, C, and H (there is only one such). Moreover, the two circles intersect in two points—A and H.

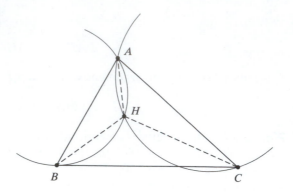

If we didn't know *H* ahead of time, maybe we could have constructed it from the two circles. What we need then is a description of the two circles that depends only on the points *A*, *B*, and *C* and independent of *H*. First of all, in one circle angle *AHB* is inscribed and has measure 120°. In the other circle, angle *AHC* is inscribed and has measure 120°. The little voice from the past again whispers, "Inscribed angles in circles: I remember studying something like that. What was it?" It might be worth taking some time out for review.

✳ *Inscribed Angles in a Circle: A Pause for Review*

In Chapter 2, in the discussion of Archimedes finding the volume of a sphere, we encountered angles inscribed in a circle. There, the angles were inscribed in a semicircle; moreover, all the angles were right angles! Could it be true that all the angles inscribed in a given arc have the same measure? (If that were the case, then any angle inscribed in arc *AB* of the circle above would measure 120°.) (You might want to use Geometer's Sketchpad to try this out.)

Here is a theorem that you might remember from previous experiences with geometry:

Theorem 3. The measure of an angle inscribed in an arc of a circle is half the measure of the central angle subtending the same arc.

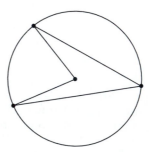

This theorem jibes with what we know about right angles inscribed in a semicircle. There, the central angle is a straight angle with measure 180°. (You might want to use Geometer's Sketchpad to verify other instances of this theorem.)

Furthermore, a corollary of the theorem is the fact that all angles inscribed in a given arc have the same measure. Also, it tells us something about the circle we're interested in for our problem: It's a circle through A and B having center X such that angle AXB has measure 120°.

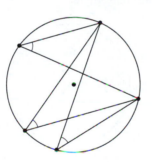 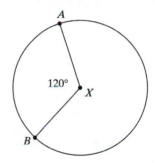

While thinking how we might find X (given A and B), let's prove one case of the theorem. Consider a circle with center Y and suppose that D, E, and F are points (in that order) lying within a semicircle of the circle. The inscribed angle we are interested in is $\alpha = \angle DEF$. The central angle is $\beta = \angle DYF$. (This is the case where the central angle is greater than a straight angle.)

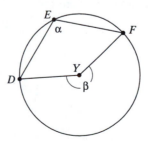

Draw the diameter through E. This touches the circle at another point G and divides angle α into α_1 and α_2 and angle β into β_1 and β_2.

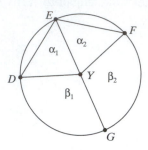

Angle β_1 being an exterior angle of triangle DEY is equal to the sum of the angles at D and E. But triangle DEY being isosceles has its two base angles equal. Thus, $\beta_1 = 2\alpha_1$. Similarly, $\beta_2 = 2\alpha_2$. Thus, $\beta = \beta_1 + \beta_2 = 2\alpha_1 + 2\alpha_2 = 2\alpha$. This proves this case of the theorem.

Problem 23

Prove the other case of the theorem, namely, when the central angle measures less than or equal to a straight angle.

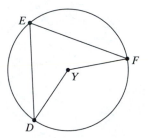

Construction of the Special Point, Concluded.   We could create the point H (which minimizes the sum of the distances in the case when the solution is not one of the three cities) by first creating two circles mentioned earlier (which intersect in A and H). One circle passes through A and B and has the property that an angle inscribed in the short arc AB is a 120° angle. We need a third point to determine the circle. By the theorem above, an angle inscribed in the long arc AB of this circle would have measure 60°.

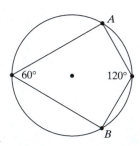

If we were able to create a point D such that $\angle ADB$ is a 60° angle, then the circle through points A, B, and D would be the one we want. It's easy to construct D: build an equilateral triangle having base AB; the third vertex will be a point D with the desired property.

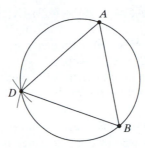

Similarly, we can create the appropriate circle passing through points A and C. The two circles will meet in a point H having the property that $\angle AHB$, $\angle AHC$, and $\angle BHC$ are all 120° angles.

Problem for Right Now!

(Continuation of the earlier Classroom Experiment) Bring out the large triangles you made earlier in groups. Carry out the construction above to create the point H. Compare this point with the other "magic" points from the point of view of sum of distances to the vertices.

✸ *The Three Cities Problem, Concluded*

We know that if one angle of triangle ABC has measure 120° or greater, then one of the cities is the place H for which $AH + BH + CH$ is smallest (problem 22). We know that if every angle has measure less than 120°, then we can construct the "120° point" N interior to the triangle. Is N always the point we're looking for? That is, is it always better than any city?

So, let's suppose that every angle of ABC has measure less than 120°. Let's try to show that $AN + BN + CN < AB + AC$. The latter is the length of the optical fiber if the solution point happened to be at city A. (If our argument works for city A, then it should work for B and C as well.)

To show this, extend BN through N and CN through N. Drop perpendiculars from A to both of these lines to obtain points D and E, respectively.

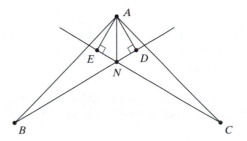

Now triangles *AND* and *ANE* are 30-60-90 triangles so that $ND = (1/2)AN$ and $NE = (1/2)AN$. Furthermore, $BD < AB$ (the hypothenuse of a right triangle is bigger than any leg) and, similarly, $CE < AC$. Also, $BD = BN + ND = BN + (1/2)AN$ and $CE = CN + NE = CN + (1/2)AN$. Thus,

$$BN + AN + CN = [BN + (1/2)AN] + [CN + (1/2)AN] = BD + CE < AB + AC$$

So the point *N* is better than city *A*. By a similar argument we can show that *N* is better than *B* or *C* as well.

Thus if all the angles of *ABC* measure less than 120° the best point is the "120° point." This solves the problem!

Problem 24
(a) Take a point *P* interior to an equilateral triangle *UVW*. Drop a perpendicular from *P* to each of the sides, meeting them in points *A*, *B*, and *C*. Show that the sum $PB + PA + PC$ is equal to an altitude of *UVW*. (b) An alternative solution to the three cities problem involves starting with cities at *A*, *B*, and *C* and then constructing the equilateral triangle *UVW*. Show how to do this.

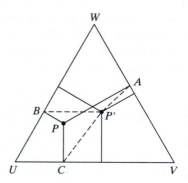

Problem 25
Here is a generalization of the Three Cities Problem to three dimensions: Take four points—*A*, *B*, *C*, and *D*—in space. Find a fifth point *P* such that the sum $AP + BP + CP + DP$ is smallest.

Problem 26
Where should four cities locate an optical fiber junction so that the sum of the distances from the junction to the four cities is a minimum? You might want to use Geometer's Sketchpad to help you with this problem.

Problem 27
Three cities are planning to build an airport. Where should they build the airport if it is to be the same distance to each of the three cities? How does the solution to this problem compare, practically, to the solution where the sum of the distances is to be minimized?

Problem 28
Four cities *A*, *B*, *C*, *D* located at the vertices of a rectangle are planning to build an optical fiber network connecting the four cities. Obviously, they want the

total length of the network they construct to be as small as possible. Below are some pictures of different kinds of networks to investigate. (Geometer's Sketchpad might be useful in your investigations.)

(a) Connect the cities directly. What is best possibility here?

(b) Connect the cities to a fifth point *P*. Where is (are) the best place(s) for *P*?

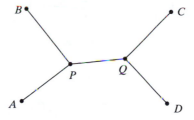

(c) Connect the cities to two additional points *P* and *Q*. Where is (are) the best place(s) for *P* and *Q*?

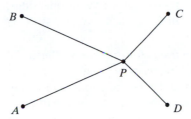

(d) Are there other possible networks connecting the four cities? (How about adding **three** more points?)

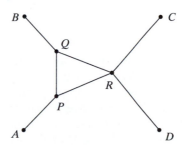

(e) Use your conclusions from (a), (b), (c), and (d) to construct the network with the least length.

Problem 29

Take four cities A, B, C, and D. Construct the network of optical fibers that connects them of least length. (Hint: Your work with problem 28 should be helpful. Also, you may want to break the problem into several cases.)

Problem 30

Suppose n points x_1, x_2, . . . , x_n are on a line in that order. Find a point P on the line so that the sum of the distances from P to the n points is a minimum. (You might want to use Geometer's Sketchpad to help you explore the possibilities.)

Problem 31

(a) Of all triangles having fixed area A and fixed length of one side, which one has the sum of the remaining two sides smallest? (Hint: milkmaid problem.)
(b) Of all triangles having one side fixed and the sum of the other two sides fixed, which one has the largest area?

Problem 32

An apparatus for drawing an "oval." A cabinetmaker friend of mine has a device for making wooden, oval picture frames. Where we attach a pencil at P on the following device, he connects a router, which then carves out the oval. Can you say more about the oval created? (It turns out that the device is called the Trammel of Archimedes. From Cundy and Rollett (1961):

Four triangular pieces are bolted firmly to a base-plate which rests on the paper. They form the walls of two slots at right angles in which the sliders A and B and run. *PAB* is a slotted arm which can be screwed to the sliders at A and B in such a way that it is free to rotate on the slider, but not to slip along the slot AB. The sliders must be longer than the width of the slots at O to ensure free travel across the opening."

By adjusting the screws, you can make "ovals" of different sizes.)

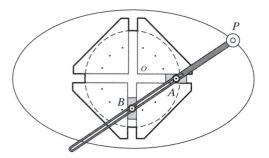

Problem 33

My daughter Katie took a mechanical drawing course in which she learned the following method for drawing an ellipse. Draw rhombus $ABCD$ with 120° angles at vertices A and C. Drop perpendiculars from A to BC and CD hitting those lines at E and F, respectively. Similarly, drop perpendiculars from C to AB and AD hitting them at G and H, respectively. Line AE intersects line CG at P and line AF intersects line CH at Q. Now you're ready to construct the "ellipse."

It's made up of four arcs of circles. With center at P and radius PG, construct the short arc of a circle joining G with E. Similarly, with center at Q and radius QF construct the short arc of a circle joining F with H. Next, with C as center and radius CG, construct the short arc of a circle joining G with H. Finally, with A as center and radius AE, construct the short arc of a circle joining E with F. This process should give you a curve passing through G, E, F, and H. What do you think about this curve? Is it an ellipse? What are its foci?

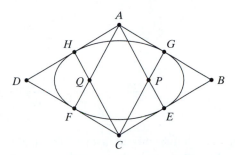

❋ *Full Circle*

The Speaker Wire Problem got us thinking about finding shortest paths in unusual situations—where the solution is not given to us on a silver platter. The first main event of the chapter was solving the Problem of the Milkmaid, River, and Cow. This problem has a clever solution (suggested by the solution to the Speaker Wire Problem: turn it into a straight line problem). The solution, in turn, has an unexpected series of consequences: First, it explains the behavior of mirrors, then it explains the reflection properties of an ellipse, and finally (after a little more discussion) it explains the reflection properties of a parabola. The second main event was solving the Three Cities Problem. The solution to the Milkmaid, River, and Cow Problem gave us a partial solution to the Three Cities Problem: One could deduce from the former that if there were a "special" point in triangle ABC, then the point would have to satisfy certain properties. (The key word in that sentence is "if.") Next, we showed how to construct a point satisfying those certain properties. Finally, we showed that if all the angles of triangle ABC are less than 120°, then the point just constructed is a special point. These three steps completed the solution to the problem.

❋ *Notes*

The Speaker Wiring Problem and the numbered problems following it barely touch on a much larger problem: Given two points on a surface, find a path of shortest distance joining them. If the surface is "smooth," then such a path is called a **geodesic**. The study of such paths was first undertaken in 1697 by John Bernoulli, whose main concern was with geodesics on the surface of the earth. The general problem was undertaken in earnest by Gauss in his definitive paper of 1827 "General Investigations of Curved Surfaces". Their study now forms a part of *Differential Geometry*. (See Coxeter, 1969, p. 366f; Hilbert and Cohn-Vossen, 1952, p. 220f; O'Neill, 1966, pp. 228–232, 236–363; and Henderson, 1996, pp. 14, 34, 35, 175, 176.)

We derived a property of flat mirrors from the solution to the Milkmaid, the River, and the Cow Problem. Historically, the flat mirror problem and its solution came first. In his *Optics*, Euclid states "light travels through space along straight lines." Also, in his *Catoptrica* ("theory of mirrors"), he proved that light traveling from P to a mirror and then to Q takes the path for which $\angle 1 = \angle 2$.

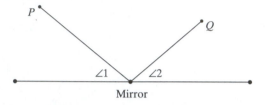

Heron (ca. A.D. 100) proved that the latter path is shorter than any other straight line path from P to mirror to Q. This property of flat mirrors is frequently derived from Fermat's Principle of Least Time (formulated in 1657): Light always takes the path requiring least time.

The Greek definition of a conic section as a cross section of a cone differs from the analytical geometry one that most of us learned. The two definitions can be reconciled, such as in the book by Jennings. The reflection properties of the ellipse were proved by Apollonius (250–175 B.C.) in his *Conics*. The reflection properties of the parabola were proved by Diocles (200 B.C.) in his treatise *On Burning Mirrors*. (See Katz, p. 118f, for this argument.) Archimedes (287–212 B.C.) seemed to know about the reflection properties of the parabola

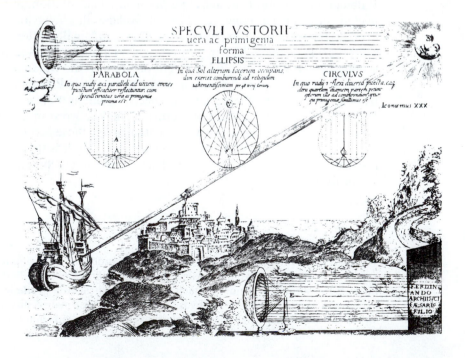

in constructing his "burning mirrors" to incinerate the Roman fleet which was laying siege to his (Greek) city of Syracuse in present-day Sicily.

In 1822, the French mathematician Germinal Dandelin proved the following interesting theorem: If two spheres are inscribed in a circular cone so that they are tangent to a given plane cutting the cone in a conic section, the points of contact of the spheres with the plane are the foci of the conic section. For more on these *Dandelin spheres*, see Courant and Robbins (p. 200f) and Jennings (p. 86f).

The Three Cities Problem was first solved by Cavalieri in 1647. Not knowing Cavalieri's result, Jacob Steiner at the University of Berlin solved it in 1842 and generalized the problem to *n* cities, called a **problem of Steiner type**. (See problems 28 and 29 in the text.) The solution to the Three Cities Problem explains certain properties of soap bubbles. Soap bubbles are used, in turn, to investigate solutions to the *n*-City Problem. (See Courant and Robbins, 1973, p. 385f.) An alternative method of solution to the Three Cities Problem can be found in Niven (p. 158f). Steinhaus (p. 119f) gives a mechanical solution to the Three Cities Problem and generalizes it to a "weighted" Three Cities Problem.

Problems of Steiner type have been used to construct telephone, pipeline, and roadway networks and, most recently, to design electronic integrated circuits in which the networks are rectilinear. See the article by Bern and Graham (1989) for more details.

THE FAR SIDE

"Well, lemme think. ...You've stumped me, son. Most folks only wanna know how to go the other way"

✳ References

Bern, M. W., and R. L. Graham, "The shortest network problem." *Scientific American*, January 1989, pp. 84–89.

Courant, R., and H. Robbins, *What is Mathematics? An elementary approach to ideas*. New York: Oxford University Press, 1973.

Coxeter, H. S. M., *Introduction to geometry*. New York: Wiley, 1969.

Cundy, H. M., and A. P. Rollett, *Mathematical models*. Oxford: Clarendon Press, 1961.

Henderson, David W., *Experiencing geometry on plane and sphere*. Upper Saddle Run, NJ: Prentice Hall, 1996.

Hilbert, David, and S. Cohn-Vossen, *Geometry and the imagination*. New York: Chelsea Publishing Co., 1952.

Hildebrandt, S., and A. Tromba, *Mathematics and optimal form*. New York: Scientific American Library: Dist. by W. H. Freeman, 1985.

Hughes Hallett, D., et al., *Calculus*. New York: Wiley, 1994.

Isenberg, Cyril, *The science of soap films and soap bubbles*. Clevedon: Tieto Ltd., 1978.

Jennings, George A., *Modern geometry with applications*. New York: Springer-Verlag, 1994.

Kappraff, Jay, *Connections: the geometric bridge between art and science*. New York: McGraw-Hill, 1991.

Katz, Victor, *A history of mathematics: An introduction*. New York: Harper Collins, 1993.

Moore, Charles G., "Contemporary conic sections," *CONSORTIUM* (Newsletter of COMAP), no.20, November 1986, p. 2f.

Niven, Ivan, *Maxima and minima without calculus*. Washington D.C.: Mathematical Association of America, 1981.

O'Neill, Barrett, *Elementary Differential Geometry*. New York: Academic Press, 1966.

Pólya, George, *Induction and analogy in mathematics, Vol I of Mathematics and plausible reasoning*. Princeton, NJ: Princeton University Press, 1954.

Steen, Lynn, ed., *For all practical purposes: Introduction to contemporary mathematics*. New York: W. H. Freeman, 1988.

Steinhaus, Hugo, *Mathematical snapshots*. New York: Oxford University Press, 1950.

Stevens, Peter S., *Patterns in nature*. Boston: Little, Brown, 1974.

Thompson, D'Arcy Wentworth, *On growth and form*. Cambridge: Cambridge University Press, 1961.

5 KALEIDOSCOPES

When I look in my bathroom mirror, I am interested in my reflection, so that I can shave, comb my hair, or put on a tie.

On the other hand, an interior decorator uses a mirror as a device for creating space. He is interested not just in the reflection of the room in the mirror, but also the "doubled" room made up of the real room and its mirror reflection. The mirror gives the illusion that the space of the room has been doubled.

With a kaleidoscope the situation is similar. There are mirrors (or maybe just a single mirror) and a real object, a "design." This design plus reflected images of it in the mirrors make up a neat overall design. It is this overall design that interests us.

As a simple instance of this, recall the example from Chapter 4. We started with the original design below.

Original design

A single mirror is placed perpendicular to a table, and the original design is placed in front of it on the table.

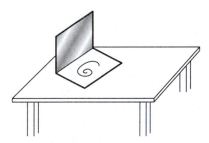

When you look down at the mirror and design from above, you will see the following overall design made up of the original and its reflection.

179

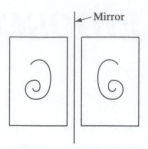

**Problem for
Right Now!**
Try this out with your own mirror and design. When you look down on mirror
and design, you want your line of sight to be in front of the mirror.

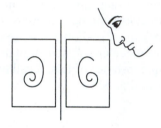

Normally a kaleidoscope contains more than one mirror. The one-mirror kaleidoscope, though useful for interior decorating, is not very interesting from the point
of view of the design it creates. The purpose of this chapter is to investigate what
combinations of more than one mirror will produce a good kaleidoscope. Of course,
we will have to decide what we mean by "good" and we will be interested in
describing the overall design each combination produces. We will start with two
mirrors and work our way up.

✸ Two-Mirror Kaleidoscopes

In this section you should have a pair of mirrors so that you can carry out for
yourself the activities described. The ideal setup would be to mount a square, high-
quality mirror 6″ × 6″ on a block of wood 6″ × 6″ × 2″. Make sure that the edges
are ground so that they're not sharp. Mount each mirror so that when the mirror-
with-block is placed on a flat table, the mirror will be perpendicular to the table
and flush with it. Make two of these.

Start with a mirror perpendicular to a table, as in the previous example, and
add a second mirror also perpendicular to the table. Place edges of the two mirrors
together so that their lines of intersection with the table form a 90° angle. Looking
down at the mirrors on the table, you should see the following.

Place a design on the table "between" the two mirrors—like this:

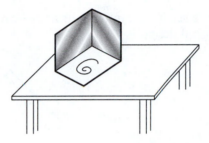

As you look down at the mirrors and the design, you should see something like the following figure lying flat on the table.

Mirrors

This neat overall design lies in the plane of the table. How can we account for it? Let's dissect the overall design. Label the parts *A*, *B*, *C*, and *D* and the mirrors 1 and 2.

Now let's see if we can reconstruct the overall design, bit by bit. First, there is *A*, the original design.

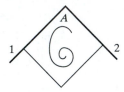

Second, there is *B*, the reflection of *A* in mirror 1.

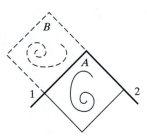

Third, there is *C*, the reflection of *A* in mirror 2.

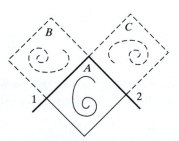

Fourth, there is D, the reflection of B in mirror 2.

(What about the reflection of C in mirror 1?)

Let's have a closer look at how this secondary reflection D is created. Pick a point P in D. You can look at P either through mirror 1 or through mirror 2. Suppose first that you look at P through mirror 1 as in the diagram below.

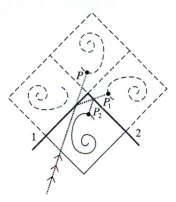

The point P is the reflection of point P_1 in region C through mirror 1, which in turn is the reflection of point P_2 on the original design.

Now look at P through mirror 2, as in the following diagram.

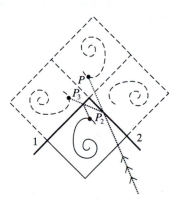

The point P in this case is the reflection of point P_3 in region B through mirror 2, which in turn is the reflection of point P_2 on the original design. Neat!

What has happened? If we look through mirror 1, then what we see of D is a reflection of C in mirror 1. If we look through mirror 2, then what we see of D is a reflection of B in mirror 2. Typically, our view of D is a little bit of both: The vertical line where the mirrors join dissect D into two pieces, the part on the left is a reflection of a part of C in mirror 1, the part on the right is a reflection of a part of B in mirror 2.

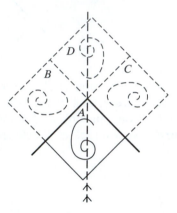

Let's consider a second example. Again take two mirrors perpendicular to a table. This time, place the two mirrors so that their traces on the table form a 60° angle.

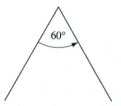

To make it easy to follow the argument, here are copies of a design to place between the two mirrors. Make a copy of the page and cut out the designs.

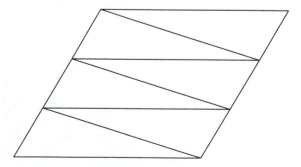

Next, place a design on the table "between" the two mirrors.

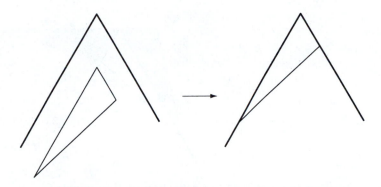

Before drawing in the overall design, let's see if we can predict what it will be by "building up" the reflections one at a time. As before, label the mirrors 1 and 2. The first part of the overall design is *A*, the original design.

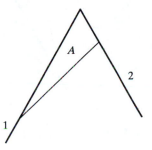

Second, there is *B*, the reflection of *A* in mirror 1.

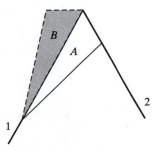

Third, there is *C*, the reflection of *A* in mirror 2.

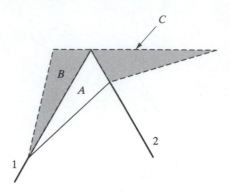

Fourth, there is *D*, the reflection of *B* in mirror 2.

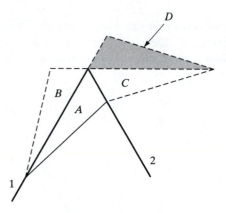

Fifth, there is *E*, the reflection of *C* in mirror 1.

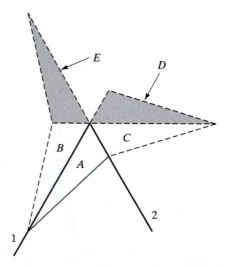

Sixth, there is *F*, the reflection of *E* in mirror 2.

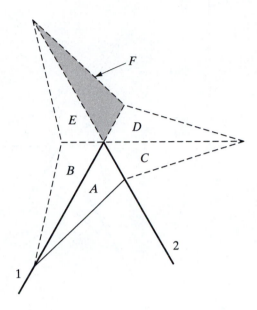

(What about the reflection of *D* in mirror 1?)

Again, choose point *P* in *F*. Let's see how it has been created by the mirrors. First, look at *P* through mirror 1. In this case, *P* is the reflection of point *P'* in region *D*.

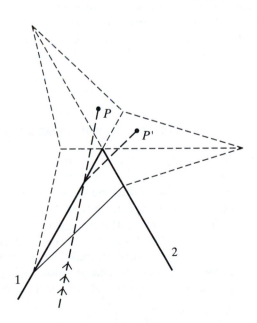

Second, look at P through mirror 2. In this case, P is the reflection of point P'' in region E.

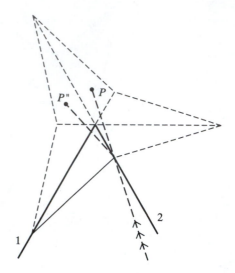

Thus, if you look through mirror 1, F is a reflection of D in mirror 1; if you look through mirror 2, F is a reflection of E in mirror 2. Typically, the view of F is a little bit of both: to the left of the vertical line we see a reflection of part of D in mirror 1, to the right we see a reflection of part of E in mirror 2.

Two mirrors set at 90° or 60° in the previous examples with an design placed between them create a pleasing **overall** design. This pleasing total design is what a kaleidoscope is all about. Will other angles similarly create pleasing designs? The following experiment begins to answer this question. Carry out this experiment before reading further.

Experiment 1 with Mirrors

(You might want to work with a partner in carrying out this experiment.)

1. Make a copy of the design below. Take two "block" mirrors, as described earlier. Place one of the mirrors along the left-hand edge of the design, as indicated. Keep this one mirror fixed. Place the other mirror along the top of the design and so that the ends of the two mirrors meet at the point indicated. Keeping the two mirrors meeting at this point, move the second mirror so that the angle between the two mirrors changes. (See diagrams below.) As you do this, look in the mirrors at the **overall design**. (For each angle a piece of the design below will play the role that the wedge played for the mirrors placed at 60° earlier.) What do you see?

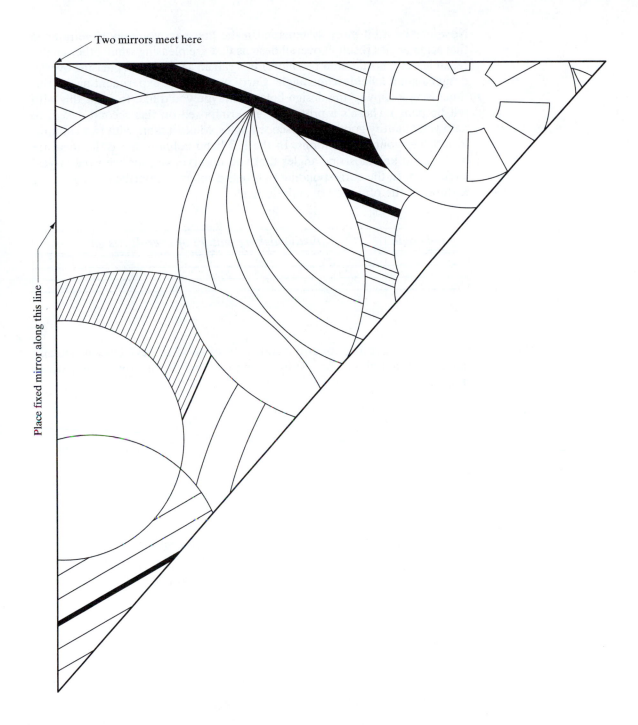

Two mirrors meet here

Place fixed mirror along this line

189

2. Now let's be a bit more systematic. In the first part you may have noticed that some angles result in overall designs that are pleasing while other angles result in designs that aren't so nice. Let's call an angle with a pleasing overall design a **good kaleidoscope angle**. Carry out the first experiment again, this time using a copy of the design below. (Maybe you'd like to color this with felt-tip pens.) There's a protractor superimposed on this second design to help you quantify good kaleidoscope angles. Make a table with two columns to tabulate your observations. In the left-hand column, mark the measure of the good kaleidoscope angles that you observe, starting with the largest angles first. In the corresponding right-hand column, describe distinguishing features of the observed overall design.

good angle	*distinguishing feature of overall design*
. . .	. . . . .

Can you say anything about the collection of good angles? Do you see any patterns? What about the distinguishing features themselves? Any pattern in them?

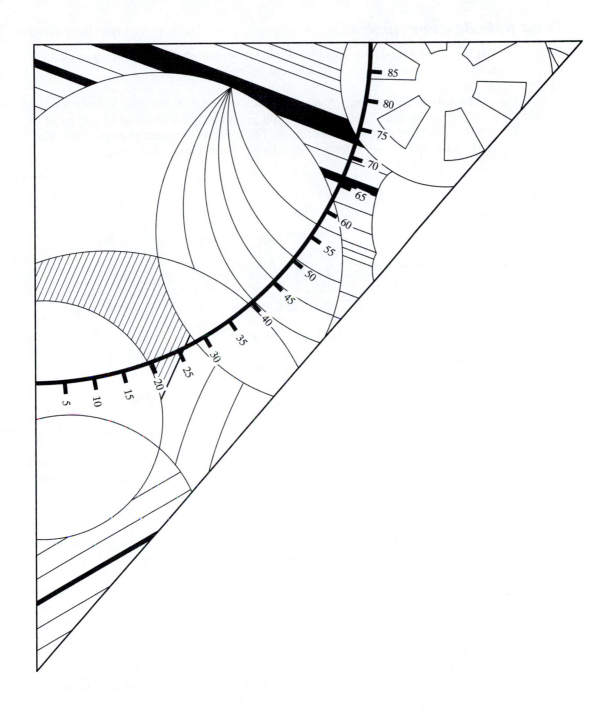

✸ Good Kaleidoscope Angles

Experiment 1 with mirrors could result in many characterizations of a good kaleidoscope angle. One of these might be the following.

Provisional Definition: A *good kaleidoscope angle* is one which produces a star as overall design. The star has n "full" points, each consisting of a copy of the original wedge plus its mirror image. The n points fit together at the point where the two mirrors meet, without overlap, to form the overall design.

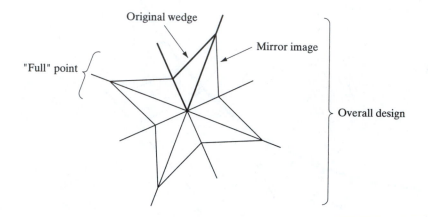

Could we use this provisional definition of good kaleidoscope angle to tell, knowing only the measure of an angle and nothing else, whether or not the angle is a good kaleidoscope angle? Let's see.

Take a good kaleidoscope angle of measure α. Thus, the angle of the wedge at the vertex where the two mirrors meet is α. The angle of its mirror image at this vertex is also α. So the angle of the star point at this vertex is 2α. If the star has n points and the points fit around the vertex without overlapping, then $n \times 2\alpha = 2\pi$. Thus, $\alpha = \pi/n$. In degrees, $\alpha = 180°/n$.

Thus, we have shown that, if α is a good kaleidoscope angle (according to our provisional definition), then $\alpha = \pi/n$ for some whole number n. Candidates for good kaleidoscope angles are therefore $\pi/2, \pi/3, \pi/4, \pi/5, \ldots$ or, in degrees, $90°, 60°, 45°, 36°, \ldots$. How do these values jibe with the good angles that appeared in the left-hand column of your table?

We know that every good kaleidoscope angle must be of the form π/n. What about the converse to this statement? For every positive whole number n, is $\alpha = \pi/n$ a good kaleidoscope angle? Out of a wedge of angle π/n we can form a "point" of angle $2\pi/n$. Then we can take n of these points to form an n-pointed star.

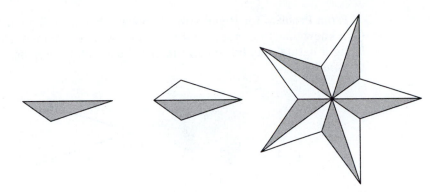

Can this *n*-pointed star be the overall design for two mirrors set at π/n? Try the following exercise.

Problem for Right Now!

On the five-pointed star below, place a pair of mirrors perpendicular to the paper along the lines indicated. Look in the mirrors. What do you see?

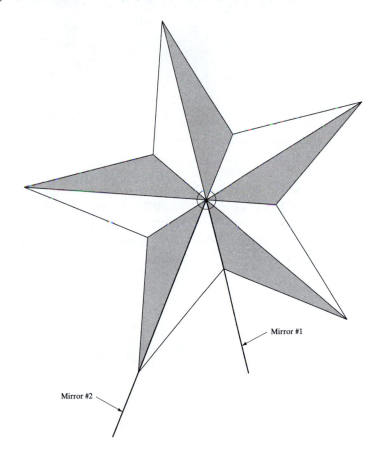

From Problem for Right Now! it appears that $\pi/5$ is a good kaleidoscope angle. (We knew that!) Let's see what we can do with a general angle of the form π/n, n being a natural number greater than 1. First, call the region behind both mirrors the *shadow*.

Second, create the overall design by adding one image of the original design at a time, letting the images spread out from the region between the two mirrors. Before reaching the shadow, the mirror images of the original alternate—direct copy, reverse copy, direct copy, . . .—without overlapping.

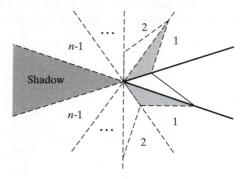

The angle between mirror 1 and the edge of the shadow is $\pi - \pi/n = (n - 1)\pi/n$. Thus, $n - 1$ copies of the original design completely fill up this region (between mirror 1 and the edge of the shadow), with no overlapping and no space left over. Call this latter region I. Similarly, $n - 1$ copies of the original design completely fill up the region between mirror 2 and the other edge of the shadow. Call this latter region II.

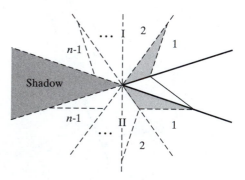

As you spread out from the original design, the last (or $n-1$st) copy C_{I} of the original design in region I is either a direct copy or a reverse copy. Similarly for the $n-1$st copy C_{II} of the original design in region II. The crucial fact is that both are direct or both are reverse.

Problem 1   Why are C_{I} and C_{II} both direct or both reverse? For what n is the $n-1$st copy direct and for what n is it reverse?

As for the examples shown earlier (90° and 60°), when you look into the mirrors at the shadow region you see a mirror image of C_{I} in mirror 2 or a mirror image of C_{II} in mirror 1 or some combination of the two. The clincher is that, since C_{I} and C_{II} are either both direct or both reverse, their images produced in the shadow **are the same** and the n-pointed star is complete. We have proved the following theorem. (You will get more out of the argument just given by carrying out the experiment and following the discussion in the next section.)

> **Theorem 1.**   An angle ϕ is a good kaleidoscope angle if and only if $\phi = \pi/n$ for some natural number $n > 1$.

The overall designs for several n are shown below.

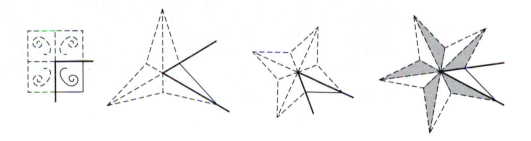

Problem 2   How does the case $n = 1$ fit into the scheme of things?

Problem 3   We began this chapter by looking at a pair of mirrors perpendicular to a table. We assumed that the traces the mirrors would make with the table were intersecting lines. What happens if the two lines do not intersect, i.e., the mirrors are parallel? Take out a pair of mirrors, set them up on a table parallel to each other. Make sure the reflecting surface of one mirror is pointed at the reflecting surface of the other. Place a design on the table between these two reflecting surfaces. Before looking in the mirrors at the overall design produced by this original and its images and the images of images (and so on), see if you can predict (as we have done with the examples above) what the overall design

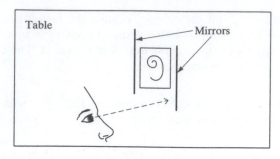

Offset the mirrors
slightly for a good view.

should be. Only after you have done that should you look into the mirrors. The
bottom line is an accounting for the overall design you see there.

✳ Bad Kaleidoscope Angles: An Analysis (Optional)

Angles of the form π/n appear to be good kaleidoscope angles. If an angle is not
of this form, the overall design will not be an n-pointed star. What will be the
overall design? Maybe angles that are not of the form are not so bad. Let's see.

**Experiment 2
with Mirrors**

(You might want to carry out this experiment with a large group of people, such
as a whole class.)

This experiment is actually a series of experiments, each similar to ones
described earlier. Each experiment corresponds to an angle which has been
drawn for you below. (The angles considered earlier were 60° and 90°.) Place
the angle flat on the table. Take a pair of mirrors and place them, perpendicular
to the table, along the sides of the angle. You want an edge of one mirror to
meet an edge of the other along a line perpendicular to the vertex of the angle.
All of this is shown in the picture below.

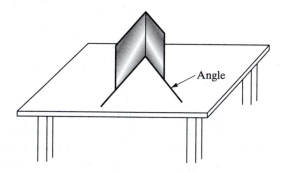

For each experiment several copies of the original design have been provided.
One copy, when cut out, should just fit between the mirrors in the given angle.
Here is a picture shown from above.

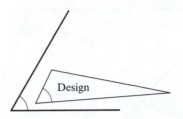

Cut out the remaining copies of the original design provided. The idea is to look at the overall design in the mirror, then, with the cutouts, try to re-create the overall design made up of the original design, plus mirror images of the original design, plus mirror images of mirror images, and so on. Use the cutout copies of the original design to carry out this re-creation much as we did earlier for angles of 90° and 60°. Thus, on a separate piece of paper, tape down a cutout copy of the original design; then tape down turned-over copies (the "reverse" design—the initial mirror images of the original) in the positions they would appear if the mirrors were actually there.

Keep doing this: Tape down the cutouts, choosing appropriately copies of the original or of its reverse and placing them in positions corresponding to mirror images of mirror images—as if the mirrors were really there. In the end, you want the design you create with the cutouts to be a copy of the overall design you see by looking into the mirrors.

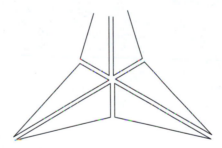

For each example, you will want to answer the following questions.
1. What do you observe? (Here is a suggestion to help you in your observations: Stare at the overall design from various positions in front of the mirrors. Direct your eye to the part of the overall design that lies in the shadow region (recall the definition of this in the previous section). Do this in two general ways: First, look through the mirror on the left; second, look through the mirror on the right. What do you see?

2. How does the design you have created compare with what you see in the mirrors?

3. Did you encounter any problems duplicating the overall mirror design using the paper cutouts?

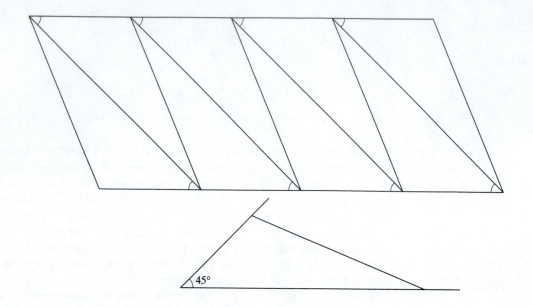

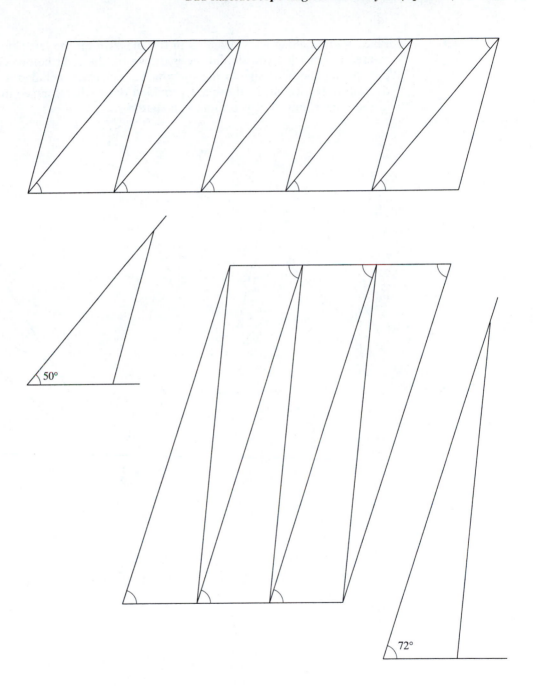

Problem 4   Place mirrors along lines *A* and *B* in the diagram on the left below. A partial re-creation of what you should see is drawn on the right below. However, the part of the overall design in the shadow has not been included. Complete the overall design. (Hint: In the space provided draw what you see in the shadow in the mirror, not what you think you **should** see!)

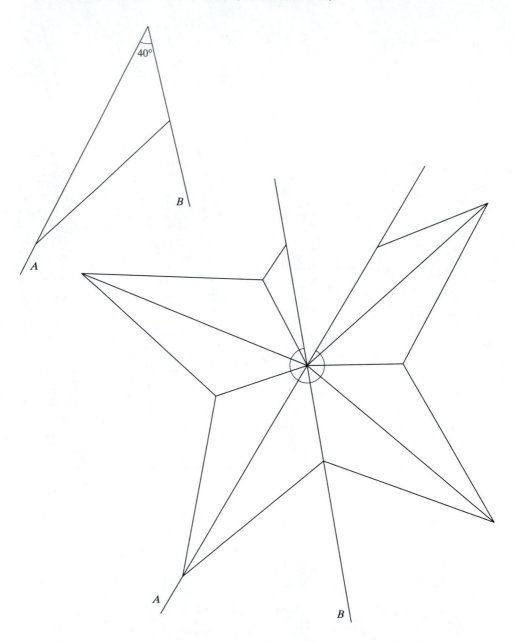

Before capitalizing on your experience with the mirrors, let's walk through another example. This time let's place the mirrors at an angle of 70°.

Let's build up the design piece by piece. In the diagram below we see *A*, the original design placed between the two mirrors.

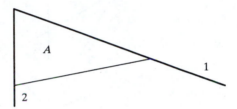

Follow the rest of the discussion with your own pair of mirrors set at 70° along the sides of the diagram just shown.

In the diagram below we add *B*, the reflection of *A* in mirror 1, and *C*, the reflection of *A* in mirror 2.

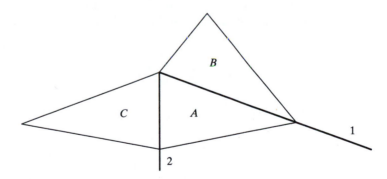

Next we add *D*, the reflection of *B* in mirror 2. This is part of what you see if you look into mirror 2. Check this out with your mirrors.

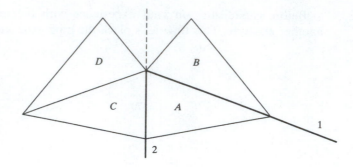

Then we add E, the reflection of C in mirror 1. This is part of what you see if you look into mirror 1. Check this out with your mirrors.

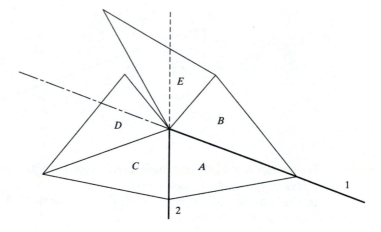

To see what else is happening, extend the line of mirror 1 so that it dissects D into two pieces, labeled D_1 and D_2 in the diagram below. (The two extended lines delineate the shadow, as discussed earlier.)

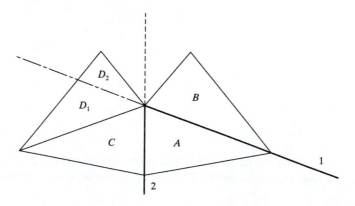

Similarly, extend the line of mirror 2 so that it dissects E into two pieces, labeled E_1 and E_2.

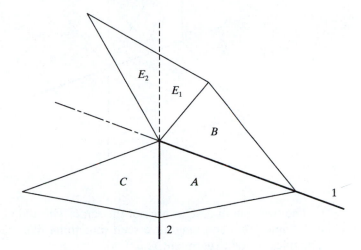

Now, if you look in mirror 2 and your line of sight is close to the line of mirror 2, then you will see F, the mirror image of E_1 in mirror 2. The overall design will look something like the following.

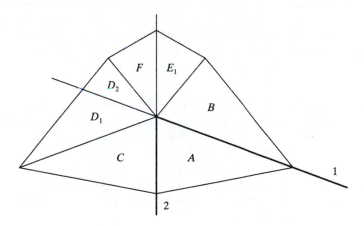

However, if you look in mirror 1 and your line of sight is close to the line of mirror 1, then you will see G, the mirror image of D_1 in mirror 1. The overall design will look something like the following.

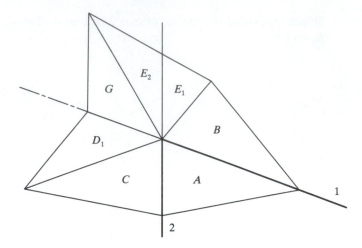

The two overall designs are not the same! This didn't happen with the examples of 90° and 60°. (You may have seen something like this occurring in some of the examples of the experiment.)

Problem 5   If you look at the overall design for 70° through mirror 1 and then through mirror 2, a portion of the overall design stays the same and a portion changes. Can you identify which is which?

Problem 6   Look into the mirrors placed at 70°. Move your eye from a sight line close to the line of mirror 2 and looking into mirror 2 to a sight line close to the line of mirror 1 and looking into mirror 1. What do you see? Concentrate particularly on the part of the design located in the shadow. What do you see now? Can you explain what you see?

There's something unsatisfactory about what you see in the 70° mirrors. Perhaps this something is due to the ambiguity in the overall design: The overall design changes as your sight line changes. Perhaps it's due to the lack of symmetry: One overall design is symmetric about the line of mirror 1 but not about the line of mirror 2; the other is symmetric about mirror 2 but not about mirror 1. For 60° and 90°, the overall design does not change with your sight line; the overall design is symmetric about the lines of both mirrors. An angle of 70° does not make a good kaleidoscope.

Which angles produce overall designs that are unambiguous and pleasing? Which angles produce overall designs that are ambiguous and not so pleasing? We'll call an example of the former an **unambiguous kaleidoscope angle** and an example of the latter an **ambiguous kaleidoscope angle.**

Problem 7   Among the examples we have discussed here and the ones you worked with in Experiment 2, which angles are unambiguous kaleidoscope angles and which ambiguous?

Let's see if we can't pin down definitively the unambiguous and ambiguous kaleidoscope angles. Let α denote the measure of the angle between a pair of mirrors that we might be investigating. Up to now two things have been important in investigating the overall design we see in the mirrors. The first is: What we see in the **shadow** of the two mirrors is the indicator of an ambiguous or unambiguous angle. The second is that the overall design is built up of direct and reverse copies of the original, starting with the original and alternating around the vertex.

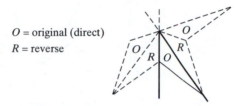

O = original (direct)

R = reverse

To investigate what we see in the shadow, we want to concentrate on the direct/reverse copies of the original design just outside the shadow. Each of these is a copy which lies entirely outside the shadow but having an edge that forms angle β with an edge of the shadow. Moreover, $0 < \beta < \alpha$. If $\alpha < 90°$, there are two of these. (See problem 10 for the case $\alpha > 90°$.)

Problem 8   Both of these copies are of the same kind: Either both are direct or both are reverse copies. Why is this? The angle β is the same for both. Why is this?

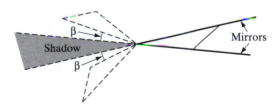

There are three cases:

1. $\beta = \alpha$.
2. $0 < \beta \leq \alpha/2$.
3. $\alpha/2 < \beta < \alpha$.

If $\alpha = 55°$, then $\beta = 15°$. This is an example of case 2. What happens here is typical of case 2. (Why is this?) A case 2 angle is therefore an ambiguous kaleidoscope angle.

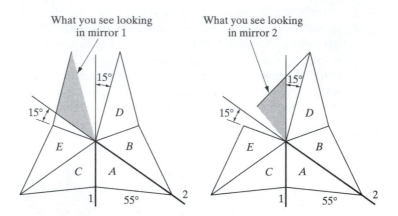

If $\alpha = 70°$, then $\beta = 40°$ and we are in case 3. Again, what happens here is typical of case 3. (Why so?) So case 3 angles are ambiguous kaleidoscope angles.

This brings us to case 1. Label X the copy in front of mirror 1 and Y the copy in front of mirror 2. The situation in this case then looks like the following.

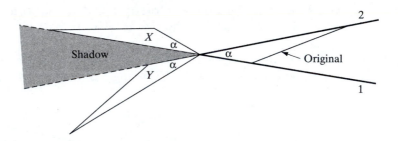

If you look in mirror 1 at the shadow, then you will see the image of X in mirror 1.

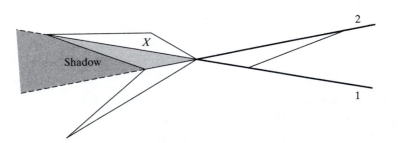

If you look in mirror 2 at the shadow, then you will see the image of Y in mirror 2.

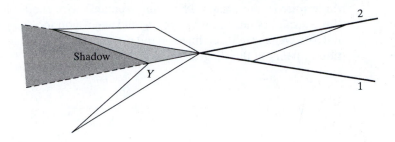

What you see in both cases is the same: The image of Y in mirror 2 and the image of X in mirror 1 are identical. There is no ambiguity. This appears to be an unambiguous kaleidoscope angle.

Problem 9   Show that, for case 1, $\alpha = \pi/n$.

Problem 10   What happens if $\alpha > 90°$ and there is only one copy of the original design outside the shadow? (This one copy is the original itself!)

The only case which yields an unambiguous angle is case 1. By problem 9, such an angle is also a good kaleidoscope angle. Thus, we have proved the following theorem.

Theorem 2. An angle is a good kaleidoscope angle if and only if it is an unambiguous kaleidoscope angle.

Problem 11   Suppose you have a pair of mirrors perpendicular to a table. Suppose that the mirrors intersect at angle α. Up to now the original design has been two-dimensional, lying flat on the table between the mirrors. What happens if you relax this last condition using a three-dimensional design? You can "see" the overall design, if you make a three-dimensional design out of a pipe cleaner. Use this as the original design. Try various angles α. What can you say about the overall design? What can you say about good and bad kaleidoscope angles? With copies of the pipe cleaner design, and a couple of good angles, construct "real" versions of the overall design you see in the mirror.

✳ Three-Mirror Kaleidoscopes I: All Mirrors Perpendicular to a Fourth Plane

We know how to place two mirrors in order to create an unambiguous overall design. So we can make good kaleidoscopes out of two mirrors. What about three mirrors? Can we make a "good" kaleidoscope out of three mirrors? What would a three-mirror kaleidoscope look like?

To get started with this problem, let's assume that we have three mirrors perpendicular to a table—just as we did with two mirrors. What are the possibilities? The three mirrors are planes whose intersections with the table are straight lines. The three straight lines are what you would see if you looked down from above on the mirrors. What can three lines on the plane of the table look like? A crude classification appears in the diagram below.

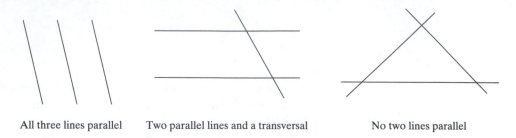

All three lines parallel   Two parallel lines and a transversal   No two lines parallel

It's hard to see how the first case could make a kaleidoscope; it would be impossible for all the silvered sides of the mirrors to face each other. We'll ignore this case. The third case (no two lines parallel) looks to be the most interesting, and we'll consider this in the next experiment. We'll leave the other case for problem 13.

Experiment 3 with Mirrors

Form a triangle with three mirrors perpendicular to the table. The view from above should look like the picture below.

The three mirrors should have their silvered sides face toward the interior of the triangle. The mirrors might extend beyond the vertices of the triangle. Place a design on the table interior to the triangle and look at the overall design created.

Think about this situation. If this triangle of mirrors is to form a good kaleidoscope, there should be no ambiguity in the overall design created. There are three places in the three-mirror configuration where two mirrors meet; these are at the vertices of the triangle. If any one of these angles is a bad kaleidoscope angle, then the three-mirror kaleidoscope will also be bad: There will be ambiguity in the overall design at that vertex. Thus, to have a good three-mirror kaleidoscope, the three angles must be good kaleidoscope angles. Experiment with different triangles, each of whose angles is a good kaleidoscope angle, to see if you can come up with good triangular kaleidoscopes. What do you find?

✳ *Good Triangular Kaleidoscopes*

A good triangular kaleidoscope must have angles which are good kaleidoscope angles. The three angles of such a kaleidoscope must have radian measures π/n, π/m and π/p for some whole numbers n, m, and p all bigger than 1.

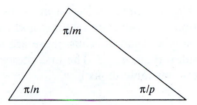

Does this narrow the possibilities for good kaleidoscope triangles? For one thing, this tells us that no angle of a good kaleidoscope triangle can be obtuse. Is there anything else we can say about these angles? Another thing we know is that the sum of these angles must be π. In other words,

$$\pi/n + \pi/m + \pi/p = \pi$$

Consequently,

$$1/n + 1/m + 1/p = 1$$

(Does this equation look familiar?)

The solutions to the latter equation will tell us about possible good kaleidoscope triangles. Let's find the solutions. First, we can assume that n, m, and p increase in size:

$$2 \leq n \leq m \leq p$$

Then, starting with $n = 2$, we can try to fill in the following table.

n	m	p
2	?	?

If $n = 2$, then $1/m + 1/p = \frac{1}{2}$. Since $n \leq m \leq p$, the first possibility for m would be 2. That doesn't work. (For one, we'd have a triangle with two right angles.) So we try $m = 3$. That forces p to be 6. So our table looks like this.

n	m	p
2	3	6

With $n = 2$, the next choice for m would be 4. This forces p to be 4, and our table looks like the following.

n	m	p
2	3	6
2	4	4

With $n = 2$, the next choice for m would be 5. This would force p to be less than 5. No good. The same argument would work for m bigger than 5: This would force p to be smaller than m. Thus, there are no more possibilities with $n = 2$.

Consider, then, $n = 3$. The first choice for m would be 3. This forces p to be 3, also. Now our table looks like

n	m	p
2	3	6
2	4	4
3	3	3

If $n = 3$ and $m > 3$, then we would have $1/p \leq 1/m < 1/3$ and

$$1/n + 1/m + 1/p = 1/3 + 1/m + 1/p < 1/3 + 1/3 + 1/3 = 1$$

No good. Similarly, if $n \geq 4$, then $1/p \leq 1/m \leq 1/n \leq 1/4$ and

$$1/n + 1/m + 1/p \leq 1/4 + 1/4 + 1/4 = 3/4$$

Again, no good. The table of possible good kaleidoscope triangles is complete as it is above.

That's astounding! There are only three possibilities, and these are the following.

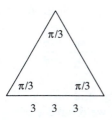

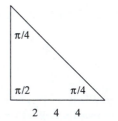

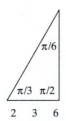

Problem 12   The three triangles just shown are the only possible triangles where the three angles are good kaleidoscope angles. For each of these, does a design placed in its interior produce an unambiguous overall design? If so, what is the design?

Problem 13   Three mirrors are perpendicular to the table. From above, two of the mirrors are parallel (their silvered parts facing each other). The third makes a transversal to these parallel lines. Are there any configurations of mirrors that create a good kaleidoscope? If so, what are the overall designs that go with them?

Problem 14   Suppose that several mirrors perpendicular to a table are arranged to form a convex polygon of four or more sides with their silvered surfaces facing toward the interior of the polygon. You want this to be a good polygonal kaleidoscope. What are the limitations, if any, on the polygon? What are the overall designs created by the good ones?

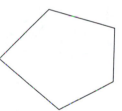

✺ *Making a Kaleidoscope*

This section is devoted to making a kaleidoscope of the type that you can buy commercially. Mirrors are inserted into a long tube. You look into one end of the tube; you turn a gadget attached to the other end to create different images. For us the tube is formed by a Pringles™ potato chip can. You might want to use something else. In fact all the materials and steps suggested below could be replaced by your own.

Materials required to make kaleidoscope:

• Pringles™ potato chip can
• Two translucent lids from Pringles™ cans

- Small items (colored beads, paper clips, . . .) to fit between the two lids
- Mirrored Mylar™ (preferably with contact backing and 5 mils thick)
- Matt board
- Colored, 1/2″ vinyl electrical tape

Tools needed:

- Hammer
- Nail setter
- Photomount glue spray can (if the Mylar™ has no contact backing)
- Heavy shears or Xacto™ knife or linoleum cutter
- Cutting board
- Sharp scissors

General description of kaleidoscope:

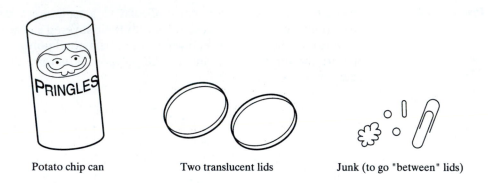

Potato chip can   Two translucent lids   Junk (to go "between" lids)

There are two possible kaleidoscopes.

1. Two-mirror kaleidoscope

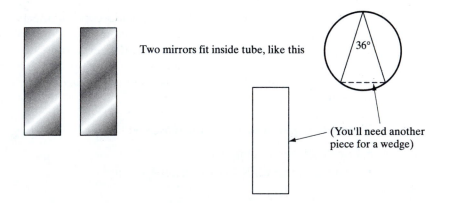

Two mirrors fit inside tube, like this

36°

(You'll need another piece for a wedge)

2. Three-mirror kaleidoscope

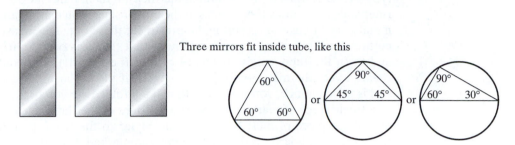

Three mirrors fit inside tube, like this

The completed kaleidoscope looks something like this:

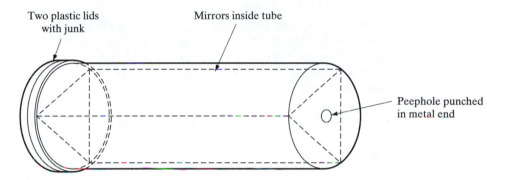

Two plastic lids with junk

Mirrors inside tube

Peephole punched in metal end

Steps to make a kaleidoscope:

- Look at the general description of the kaleidoscope above and how it is put together. Decide which type to make.
- **Make a viewing hole for the kaleidoscope.** Use a nail setter and hammer to punch a hole in the metal end of the potato chip can. For most types of kaleidoscopes, the hole should be in the middle of the circle; for some, however, you may want to offset the hole.
- **Make the part of the kaleidoscope that forms the basis for the design (this is the part you turn).** Put colored beads, etc., between the two Pringles™ lids. Use electrical tape to seal together the two lids.

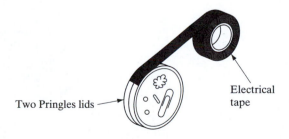

Two Pringles lids

Electrical tape

- **Cut the matt board to the shape of the mirrors.** Find the measurements for the type of kaleidoscope you want. (Some measurements are shown below.) With ruler and pencil, draw the outlines of the mirror shapes on matt board. Use knives, metal straight edges, and cutting boards to cut these out. The measurements and cutting should be carried out carefully so that the sizes of the mirrors are exact and fit in the tube snugly. If you're careful (i.e., exact!), you could use heavy shears to cut the matt board. [Alternatively, if the Mylar has a contact backing and is thick enough, you may want to make a folded version of three mirrors out of a single piece of matt board. Mark the edges of the three mirrors on the back of a single piece of matt board. Fix the Mylar to the front. (See below.) Then score the matt board on the back where you marked it. Fold to make the three-mirror configuration.]

- **Cut out a piece of Mylar.** Cut out a single piece of Mylar big enough to cover the pieces of matt board.

- **Make the mirrors.** Remove contact backing, place Mylar on paper towels on table, sticky side up. Place matt board pieces on top of Mylar. Press matt board pieces to smooth out Mylar and form good bond. (If your Mylar doesn't have a contact backing: To avoid ventilation problems you will need to go outdoors for this. With "Photo Mount Spray Adhesive" spray one side of the matt board pieces you have just cut out. Place the Mylar on these. Turn matt board over onto towel and press to smooth Mylar out.) With scissors, trim away excess Mylar from the matt board pieces. Careful! Mylar scratches easily! Make sure that no adhesive touches the part of Mylar that will form the reflective surface of your mirrors.

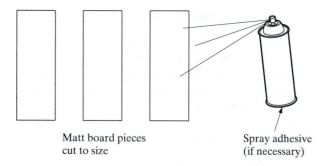

Matt board pieces
cut to size

Spray adhesive
(if necessary)

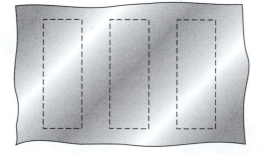

Place Mylar
on sprayed
surfaces

Turn matt board and Mylar over onto clean paper towels. Press back of matt board to smooth Mylar out and make clean bonding of Mylar with matt board.

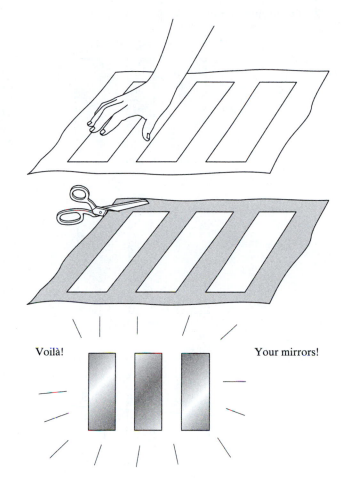

Trim away excess Mylar from matt board pieces.

Voilà!   Your mirrors!

• **Finally, assemble the kaleidoscope!** Enjoy!

Mirror measurements for different types of kaleidoscopes (cut matt board pieces to these sizes):

1. Two-mirror kaleidoscope (36° angle).

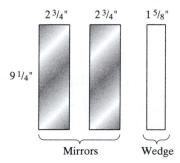

$2\,^3/_4$" $2\,^3/_4$" $1\,^5/_8$"

$9\,^1/_4$"

Mirrors   Wedge

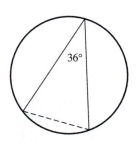

36°

2. Three-mirror kaleidoscope.

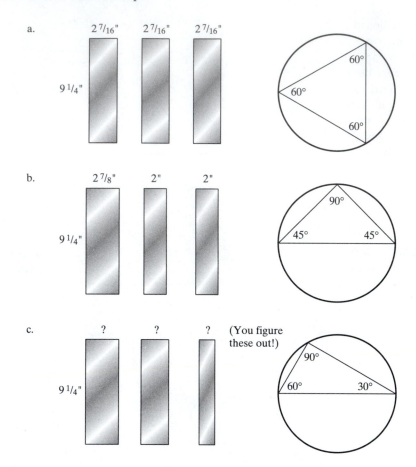

Note: The measurements here are approximate. You may have to do some trimming. After inserting the mirrors in the tube, check the angles between the mirrors with a protractor. The most successful three-mirror kaleidoscope is the 60°-60°-60°. It tolerates error best. The mirrors wedge themselves into the tube. For the other two possibilities, you may want to add glue or wedges to keep the triangle propped in position.

✳ *Three-Mirror Kaleidoscopes II: Three Mirrors Meeting in a Point*

Place a mirror on the table (silvered part facing up). On top of this place a pair of mirrors perpendicular to the table. We considered an example of this in problem 12 (b) of Chapter 4.

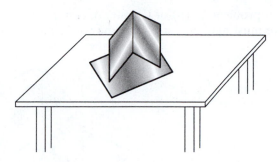

Is such a configuration of mirrors a candidate for a good kaleidoscope? We just finished looking at kaleidoscopes made up of three mirrors perpendicular to a fourth plane, the plane of the table. In this case the three mirrors meet in a point. Certainly, if the three-mirror configuration is going to make a good kaleidoscope, then the angle between any two of the mirrors must be a good kaleidoscope angle. How is the angle between a pair of planes measured in general? In the past a pair of mirrors has been perpendicular to a third plane and the angle measured has been that made by the traces of the mirrors on this third plane. This is called the **dihedral angle** between a pair of planes.

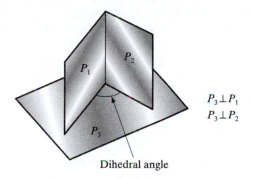

$P_3 \perp P_1$

$P_3 \perp P_2$

Dihedral angle

Thus, if three mirrors meeting at a point is to be a candidate for a good kaleidoscope, then the three dihedral angles must be good kaleidoscope angles. For the three mirrors above, the three dihedral angles are $\pi/2$, $\pi/2$, and α.

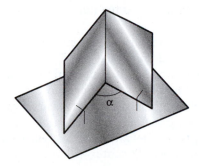

[In problem 12 (b) of Chapter 4, $\alpha = \pi/2$ also.] For the kaleidoscope to be good we must have $\alpha = \pi/n$ for some whole number $n > 1$.

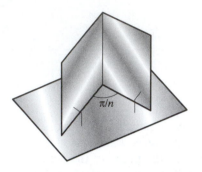

Problem 15   Suppose you have three mirrors meeting in a point such that the three dihedral angles are $\pi/2$, $\pi/2$, and π/n. You want to place a design in the mirrors and find out what will be the overall design. In the past, the original has been placed on the table between two mirrors (in the case of a two-mirror kaleidoscope) or between all three pairs of mirrors (in the case of a three-mirror triangular kaleidoscope). Where should the original design be placed in this new case? Maybe a better question would be: What should the design be? For your choice of original design, describe the overall design created by the three mirrors. (You might want to start out with $n = 2$, 3, or 4.)

In order for three mirrors meeting in a point to form a good kaleidoscope the three dihedral angles must be of the form π/n, π/m, and π/p. In the case where we had three mirrors forming a triangle perpendicular to the table, we also concluded that the dihedral angles had to be of the form π/n, π/m, and π/p. Not every combination of three such angles was possible. The three angles were also angles of a planar triangle and were therefore related by the formula

$$\pi/n + \pi/m + \pi/p = \pi$$

This equation limited the number of good kaleidoscope triangles to three possibilities.

Are there similar restrictions on the possible angles for three mirrors meeting in a point? We know that $n = 2$, $m = 2$, and $p =$ anything are possibilities. (The situation is already different from the triangular kaleidoscope where there are only three combinations.) Are there more possibilities? To find out, we turn to a sphere.

Suppose we have three mirrors meeting at a point which we assume is the center of a sphere. Each mirror, being a plane, intersects the sphere in a great circle. (In fact, this is the definition of a **great circle**—the intersection of a plane passing through the sphere's center with the surface of the sphere itself. A familiar example of a great circle on the surface of the earth is the equator; any north-south meridian (passing through North and South Poles) is half of a great circle.

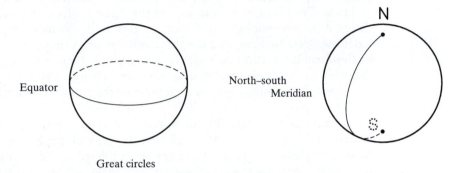

Great circles

The intersection of the three mirrors with the sphere form three great circles on the sphere. The silvered parts of the mirrors point toward the interior of a **spherical triangle** ABC on the surface of the sphere.

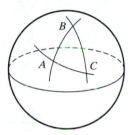

Spherical triangle ABC

The angle at A is measured in the following way. At the point A construct two tangent lines, one to one of the great circles meeting at A, the other to the other great circle. These two lines through A determine a plane, the plane tangent to the sphere at point A. On this plane, measure the angle between the two lines. This measure is also the measure of the spherical angle at A.

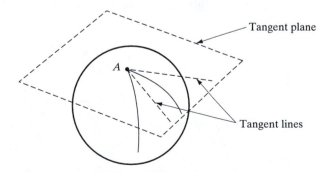

Now the tangent plane is also perpendicular to the two mirror planes determining the great circles that meet at A. The angle we have just measured on the tangent plane is also the dihedral angle between the two mirror planes.

In the Euclidean plane, the sum of the three angles of a triangle is π. This fact dictated the relationship $\pi/n + \pi/m + \pi/p = \pi$ for the three angles of a triangular kaleidoscopes. Now, in case the three mirrors meet in a point, the three dihedral angles form the vertices of a spherical triangle. Is there anything we can say about the sum of the three angles of a spherical triangle? It turns out that there is something, and it might help us to determine the possibilities for the corresponding kaleidoscopes.

Surprisingly the angle sum of a spherical triangle is intimately connected with its area. Let's see why. First, recall that the area of a sphere of radius R is $4\pi R^2$. Next, consider an angle α with vertex A on the sphere. This angle was formed by two great circles meeting at point A. The angle and two great circle halves also form a **lune** of angle α; the lune is the part of the sphere between the two meridians which form the angle. Denote the area of this lune by $L(\alpha)$.

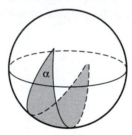

Lune of angle α and area $L(\alpha)$

Let's calculate $L(\alpha)$. First, $L(\alpha)$ is proportional to α, i.e.,

$$L(\alpha) = K\alpha$$

(Why is this?) Second, since $L(2\pi) = 4\pi R^2$, it follows that $K = 2R^2$ and therefore

$$L(\alpha) = 2\alpha R^2$$

Next, let's use this formula to figure out the area of a spherical triangle having angles **a**, **b**, and **c**.

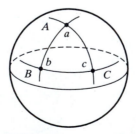

Spherical triangle of angles a, b, and c

Label the vertices at which these angles occur by the corresponding capital letters A, B, and C. The sides of this triangle are great-circle arcs AB, AC, and BC. If the great-circle arcs AB and AC are continued beyond vertices C and B, they meet again at A', antipodal to A. (A line through the center of the sphere will intersect the sphere in two points X and X', called **antipodal** points.)

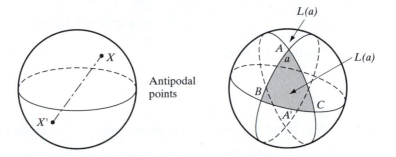

These extended arcs (from A to A') are meridians and create a lune of angle **a**. Furthermore, all points antipodal to points in this lune form another lune, also of angle **a**. Extending the sides of the spherical triangle from each of the other two vertices B and C in turn, we create two more lunes of angle **b** and two more of angle **c**. So we have six lunes in all. The sum of the areas of the six lunes is

$$2L(\mathbf{a}) + 2L(\mathbf{b}) + 2L(\mathbf{c}) = 4R^2(\mathbf{a} + \mathbf{b} + \mathbf{c})$$

Now, the six lunes together cover the entire surface of the sphere: Every point of the sphere lies in at least one of the lunes. In this covering, the spherical triangle ABC is covered three times. The same is also true of the antipodal triangle $A'B'C'$. Any point not in one of these two triangles is covered only once. (Get a sphere and check this out!) Thus,

$$4R^2(\mathbf{a} + \mathbf{b} + \mathbf{c}) = 4\pi R^2 + 4 \text{ times the area of spherical triangle } ABC$$

Consequently,

$$\text{area of } ABC = R^2(\mathbf{a} + \mathbf{b} + \mathbf{c} - \pi)$$

Since the area of the triangle must be positive, it must be that

$$\mathbf{a} + \mathbf{b} + \mathbf{c} > \pi$$

To sum up, we have the following theorem.

Theorem 3. In a spherical triangle with angles **a**, **b**, and **c**, the sum of these angles must be greater than a straight angle.

What are the consequences of this theorem for kaleidoscopes? Suppose three mirrors meeting in a point have the three dihedral angles **a**, **b**, and **c**. These are also three angles of a spherical triangle. By the theorem above, we know that **a** + **b** + **c** > π. If the configuration of mirrors is to form a good kaleidoscope, then also **a** = π/n, **b** = π/m, and **c** = π/p for some whole numbers n, m, and p. Putting all of this together, we get

$$\pi/n + \pi/m + \pi/p > \pi$$

Therefore, we have

$$1/n + 1/m + 1/p > 1$$

(Does this look familiar?)

Let's find the solutions to this inequality. We can assume first that

$$2 \le n \le m \le p$$

Next, let's try to fill in the table below. We start with $n = m = 2$.

n	m	p
2	2	?

We notice that, if $n = m = 2$, then p can be anything greater than or equal to 2. This jibes with the examples that we used to start out this section. Now let's try $n = 2$ and $m = 3$. This and the inequality $1/n + 1/m + 1/p > 1$ give us

$$1/p > 1/6$$

Thus, $p = 3$, 4 or 5.

n	m	p
2	3	p
2	3	3
2	3	4
2	3	5

Next try $n = 2$, $m \ge 4$. Since $4 \le m \le p$, it must be that

$$1/n + 1/m + 1/p \le 1/2 + 1/4 + 1/4 = 1$$

This is no good. The next case to try is $n \ge 3$. Since $3 \le n \le m \le p$, it must be that

$$1/n + 1/m + 1/p \le 1/3 + 1/3 + 1/3 = 1$$

Again, this is impossible. The previous table is complete!

**Experiment 4
with Mirrors**

We know something about the overall designs produced by three mirrors meeting at a point where the dihedral angles are $\pi/2$, $\pi/2$, and π/p—corresponding to the first row of the previous table. To get some idea of the overall designs produced by mirrors arranged so that their dihedral angles correspond to the other rows of the table, let's set up some mirrors so that we can find out. Take three mirrors labeled 1, 2, and 3 and place them on a table and perpendicular to it to form a triangle with angles $\pi/3$, $\pi/6$, and $\pi/2$, i.e., a 60°-30°-90° triangle. Looking down at the table you should see the following.

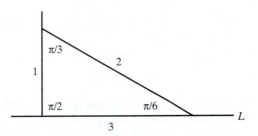

The plane of mirror 3 intersects the table in line L. The idea of this experiment is to keep mirrors 1 and 2 fixed and rotate mirror 3 about line L. The dihedral angle between mirrors 1 and 2 will remain $\pi/3$, and the dihedral angle between the planes of mirror 3 and mirror 1 will remain $\pi/2$. This is consistent with what you see in rows 2, 3, and 4 of the table: The first two dihedral angles are $\pi/2$ and $\pi/3$ for all three kaleidoscopes. However, as mirror 3 is rotated about line L the dihedral angle between the planes of mirrors 3 and 2 will change.

Start the experiment by placing mirror 3 flat on the table, like this.

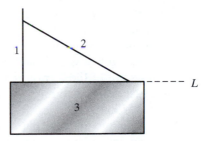

Keeping one end of mirror 3 fixed along line L, gently lift the other end and begin to slowly rotate it slowly about L. As you do this, look into the mirrors to see what you can. Try to identify those positions (and the mirror images that correspond to them) during the rotation when the dihedral angles between mirrors 2 and 3 are $\pi/3$, $\pi/4$, and $\pi/5$. What happens? (Thanks to H. S. M. Coxeter, 1969.)

Problem 16   What does the formula for the area of a spherical triangle say about the notion of similar triangles for spherical triangles?

Problem 17   We know how to build the kaleidoscopes corresponding to the first line in the table. (Problem 12 also deals with the overall designs created by them.) What about the last three lines of the table? Can these be built? What overall designs do they create?

Problem 18   (a) Find a formula for the area of a convex, spherical n-gon having interior angles equal to $\alpha_1, \alpha_2, \ldots, \alpha_n$. (b) What can you say about the sum of the interior angles of a convex, spherical n-gon?

Problem 19   What can you say about the possible good kaleidoscopes formed by n mirrors meeting in a point, $n > 3$? (The configuration of mirrors for such a kaleidoscope should look something like a pyramid with a base equal to a convex n-gon, $n > 3$.)

✸ *Full Circle*

In Chapter 4, we found out how a light ray reflects off of an object and from that how a flat mirror works. In this chapter, we exploited that knowledge in investigating what happens when there are two flat mirrors placed at an angle. We spent a lot of time trying to find out which angles produce "good" overall designs. The first hurdle was coming up with a definition of what was a good overall design. (Even before this, we had to realize that a definition was necessary!) Once we had a provisional definition, it was easy to find a lot of good angles—ones that produced good designs. We had a theorem: If $\theta = \pi/n$ (n a whole number greater than 1), then θ is a good angle, and conversely. The second, harder hurdle was to show that our provisional definition was satisfactory. In an optional section we sketched a proof that all angles not of the form $\theta = \pi/n$ produced ambiguous overall designs.

Glowing with success for two mirrors, we added a third mirror and looked for other possibilities for kaleidoscopes. We considered two cases. For the first case, the planes of the three mirrors were perpendicular to a fourth plane and intersected the latter in a triangular region. Using our earlier theorem about good angles and the fact that the angles of a triangle in the plane sum to π, we came up with a list of all possible "triangular" kaleidoscopes. For the second case, the planes of the three mirrors met in a point. This time we took a brief excursion into spherical geometry and showed that the angles of a spherical triangle sum to a quantity greater than π. Using this fact and our earlier theorem about good angles, we came up with a list of all possible "3-D" kaleidoscopes.

✻ *Notes*

The Scotsman Sir David Brewster invented the two-mirror kaleidoscope in 1813 and filed a patent for it in 1816. A crude version is one of the kaleidoscopes described in this chapter—two mirrors fitted appropriately in a tube, a peep-hole in one end of the tube, and colored items enclosed in a translucent, turning cap on the other. He coined the term kaleidoscope from the Greek kalos (beautiful), eidos (a form), and skopein (to see). Brewster is also credited with the invention of the color wheel in which red, blue, and yellow are identified as primary colors. Following Brewster's invention, kaleidoscopes were an immediate success and remained popular throughout the nineteenth century as the "great philosophical toy." You will find illustrations of several kaleidoscopes from this era as well as many modern ones in the book by Boswell (1992).

The 1819 book by Brewster is a treatise on the theory and history of the kaleidoscope in which he reports that the first published account of a primitive kaleidoscope (two mirrors hinged at an appropriate angle) was by Athanasius Kircher in 1646. Brewster noted that the angles allowed by Kircher ($360°/n$) should have been restricted to $180°/n$ (n a whole number bigger than 1).

Kaleidoscopes formed by three mirrors meeting in a point were invented by the German mathematician A. F. Möbius in 1852. The formula for the area of a spherical triangle was first published by the Flemish mathematician Albert Girard in 1629. More on spherical geometry can be found in the references, at the end of this chapter, by Henderson (1996) and Melzak (1983).

Many modern sources, such as the books by Coxeter (1969) and Kappraff (1991), discuss kaleidoscopes in the context of the notion of symmetry or the algebraic structure of a group. We will do the same thing in the next chapter. *Dihedral kaleidoscopes* is a film/video on the mathematics of kaleidoscopes using this point of view.

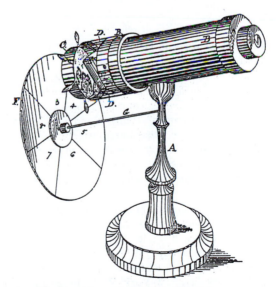

Patent illustration of the popular C.G. Bush parlor scope.

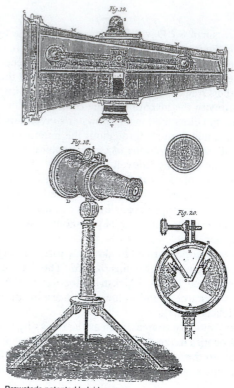

Brewster's patented kaleidoscope

❋ *References*

Ball, W. W. Rouse, and H. S. M. Coxeter, *Mathematical recreations and essays*. New York: Macmillan, 1962.

Boswell, Thom, ed., *The kaleidoscope book*. New York: Sterling, 1992.

Brewster, Sir David, *A treatise on the kaleidoscope*. Edinburgh: Constable, 1819.

Coxeter, H. S. M., *Introduction to geometry*. New York: Wiley, 1969.

"Dihedral kaleidoscopes" (film). National Science Foundation and Minnesota School of Mathematics, 1966. Distributed by International Film Board, Chicago, IL.

Henderson, David W., *Experiencing geometry on plane and sphere*. Upper Saddle River, NJ: Prentice Hall, 1996.

Kappraff, Jay, *Connections: the geometric bridge between art and science*. New York: McGraw-Hill, 1991.

Kennedy, Joe, and Diane Thomas, *Kaleidoscope math*. Palo Alto: Creative Publications, 1978.

Melzak, Z. A., *Invitation to geometry*. New York: Wiley, 1983.

O'Daffer, Phares, and Stanley R. Clemens, *Geometry: an investigative approach*. Menlo Park, CA: Addison-Wesley, 1995.

Schattschneider, Doris, *Visions of symmetry: notebooks, periodic drawings, and related work of M. C. Escher*. New York: W. H. Freeman, 1990.

Shubnikov, A. V., and V. A. Koptsik, *Symmetry in science and art*. New York: Plenum Press, 1974.

6 Symmetry

We classify shapes in the plane in many ways. There are shapes which are curves, and some which aren't. Some curves are closed, some not. Some closed curves are polygons and some not. Among polygons, there are those which have three sides, those which have four sides, ... We might want to classify shapes by the area they occupy: This shape has area 10 cm², that shape has area $\sqrt{2}$ ft², ... All of these methods of classification involve measuring length, area, or angle, or numbering the pieces of a certain kind that the shape might possess. Another way we might classify a shape is by symmetry: That shape has it, this shape doesn't. How do you tell if a shape has symmetry? And is symmetry something we ought to be discussing in a math book? Symmetry has an aesthetic, subjective feel to it, something that brings out a comment like "I don't know anything about art, but I know what I like!" To get a handle on symmetry, let's try the following exercise.

Problem for Right Now!

Get in a group with two or three other people. Consider the following shapes in the plane.

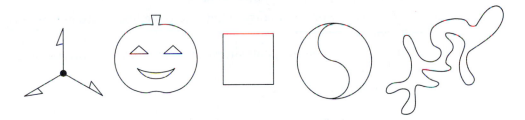

In your group discuss these questions: (1) Which of the shapes have symmetry? (2) For each shape that has symmetry, what is it about the shape that makes it symmetric?

There are lots of answers to the questions in Problem for Right Now!, and all of them correct! After all, we're mushing around, looking for a definition of "symmetry" we can agree on. One discussion might go like the following.

SARAH: The pumpkin is symmetric because it's the same on both sides.

JOHN: What do you mean by "both sides"?

SARAH: If you drew a line from top to bottom, from stem to base, then both sides of the pumpkin would be the same.

MARIA: That would be true of **this** shape and the line I've drawn: Both sides are the same. But would you call it symmetric?

SARAH: Hmm. No. I know. One side is a mirror image of the other.

JOSE: What is that supposed to mean?

SARAH: If you put a mirror right where the line is, the mirror image of one side will be the other.

JOSE: I see. Half the pumpkin plus its mirror image forms the total design, which is the whole pumpkin.

MARIA: Yes!

The discussion suggests a start to defining symmetry. Take a shape and a line *L* in the plane. Place a mirror along *L* perpendicular to the plane. Part of the shape will be in front of the reflecting surface of the mirror. Part of the shape will be in back of the reflecting surface of and covered up by the mirror. If the image in the mirror recreates the part covered up, then the shape has *mirror symmetry along line L.* Here are some examples.

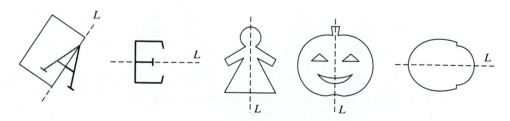

Some shapes have mirror symmetry along several lines.

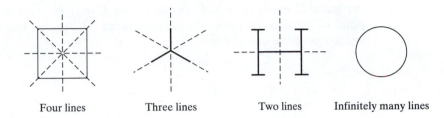

Four lines   Three lines   Two lines   Infinitely many lines

Problem 1   Find all of the lines of mirror symmetry (if any) for each shape.

✳ *Mirror Symmetry and Transformational Geometry*

There is another way to think about mirror symmetry. The mirror symmetry of a shape in the plane can be accounted for by subjecting the entire plane to a certain *transformation,* or *motion.* We know that the pumpkin face shown below has line *L* of mirror symmetry. Imagine that the line is a "spit" with which you have "speared" the plane; then, holding the spit fixed, slowly "turn the plane over." When you have done this, the pumpkin face appears as if it had not been changed. This would not have happened if the pumpkin had not had mirror symmetry in line *L.*

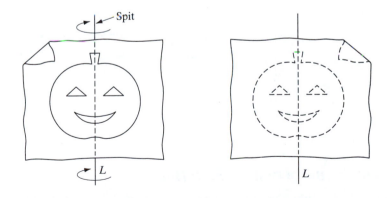

This transformation—turning the plane over using "spit" *L*—is sometimes called a *flip about L.* A shape has mirror symmetry along line *L* if the flip about *L* leaves the shape unchanged in how it looks.

A way to express the fact that the tree below has mirror symmetry along line *L* is to trace a copy of the tree (and line *L*, too) on a separate piece of paper. Flip the piece of paper over and lay the traced *L* over the original, with each traced point of the line lying over the original point. The traced tree (flipped) should fit exactly over the original tree.

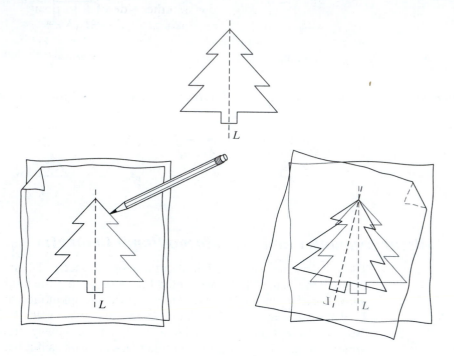

Problem 2 Use tracing (see-through) paper to make a tracing of each of the following. In each case, test for the mirror symmetry of the shape along line *L*.

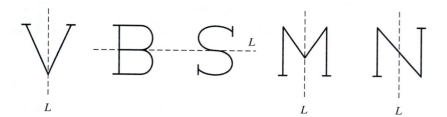

✳ *Mirror Symmetry as a Function*

The flip about line *L* suggests another way of thinking about mirror symmetry. You can interpret the flip about line *L* as a function that sends points in the plane to points in the plane. Denote the function by M_L and call it the *mirror reflection in line L*. The domain of this function will be the set of all points in the plane; its range will consist of points in the plane as well. If you input point $X \in \mathbf{R}^2$ and output point $M_L(X) \in \mathbf{R}^2$, then $M_L(X)$ is the mirror reflection of point X in a mirror placed along line L. This time you imagine that the mirror is silvered on both sides. In other words, through point X drop a perpendicular to line L. Let it meet L at point P. Extend this line on the other side of L to point Y such that $YP = XP$. The point Y is what we are looking for: $Y = M_L(X)$. Informally, we can say that M_L *sends* point X to point Y.

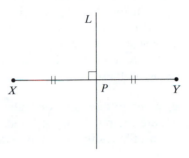

The function M_L seems to capture the results of the motion of "flip about line L." The flip about line L results in point X being moved to point Y. Or, when you flip the plane about line L, the point X gets superimposed onto Y.

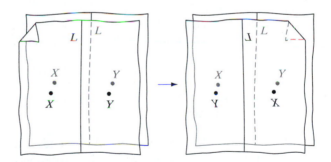

If shape S is symmetric along line L, then the flip about L leaves S looking as if it had been unchanged. (If you blink during the flip and open your eyes, you won't know anything happened to S.) How do we express this using M_L and the language of functions? Informally, we might say "M_L takes S to itself." Let's say this a bit more formally. Recall some notation about functions: Let f be a function with domain D and suppose that A is a subset of D. Then $f(A) = \{f(X): X \in A\}$. Thus, we have the following definition.

Definition: Let S be a subset of the plane and L a line in the plane. Then S has *mirror symmetry along line L* if and only if $M_L(S) = S$.

We started with something we thought was pretty fuzzy and subjective and somehow have wound up with this formal definition. It may look a little formidable. Here are some consoling words. First, there's more to the story about symmetry. The present discussion simply led us to this point. Second, this new perspective on mirror symmetry opens up all kinds of possibilities. Because functions are something we all know about from previous experience with mathematics, we get to play the "comparison game" and ask a lot of questions whose answers should give us new insight into our new function, and hence to mirror symmetry itself. When I see a new four-legged creature at the zoo, I ask: How is it the same as (or how does it differ from) other four-legged animals I know? Does it bite? Can I pet it? Does it bark, meow, moo? As I get some answers, my interest is aroused and I get to "know" the new animal. What questions can we ask of M_L? Some questions to get started with are contained in the following problems. Think of some others while you're working on these.

Problem 3 (a) Is M_L a one-to-one function? That is, if X_1 and X_2 are two distinct points in the plane, are the points $M_L(X_1)$ and $M_L(X_2)$ distinct? (b) What is the range of the function M_L? Justify your answers.

Problem 4 What does M_L do to lines? That is, if L' is a line, can you say anything about the set $M_L(L')$? (A special case here is when $L' = L$.) Justify your answers.

Here are some properties of the function M_L that it might not have occurred to you to ask about.

Property 1. M_L preserves distances. This means that, if $d(X, Y)$ denotes the distance between points X and Y in the plane, then

$$d(X, Y) = d(M_L(X), M_L(Y)).$$

In words, the distance between two points is the same as the distance between their mirror images.

Property 2. M_L reverses orientation. In other words, if S is a pair of coordinate axes in which the y-axis is reached by rotating the x-axis 90° **counterclockwise,** then $M_L(S)$ is a pair of coordinate axes in which the y-axis is reached by rotating the x-axis 90° **clockwise.** In other words, M_L reverses the roles of clockwise and counterclockwise.

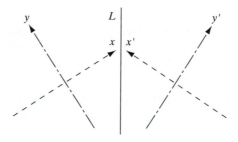

Problem 5   Show that property 1 is true. Pay particular attention to the case when X and Y are on different sides of L.

Problem 6   Suppose that A, B, C are three points in the plane and $A' = M_L(A)$, $B' = M_L(B)$, $C' = M_L(C)$. How do the two triangles ABC and $A'B'C'$ compare?

Problem 7   If C is a circle of radius r, what can you say about $M_L(C)$?

Problem 8   Use Geometer's Sketchpad to investigate the properties of M_L and what it does to lines, circles, triangles, and orientation.

When we looked earlier at shapes with mirror symmetry, we noticed that some had more than one line of mirror symmetry. What else can we say when such a situation occurs?

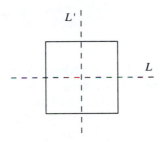

Suppose that for shape S there are distinct lines L and L' such that S has mirror symmetry along L and along L'. From the functional point of view, this means that $M_L(S) = S$ and $M_{L'}(S) = S$. Thus, the flip about L leaves S apparently unchanged; the flip about L' also leaves S apparently unchanged. Furthermore, if we transform the plane twice in succession: Flip about L followed by flip about L', then S will appear not to have changed. In terms of functions, the result of the succession of two flips is expressed by the composition of M_L with $M_{L'}$, i.e.,

$$M_{L'} \circ M_L$$

In other words, if X is a point in the plane and Y is where X goes after the two flips, then

$$Y = M_{L'} \circ M_L(X) = M_{L'}(M_L(X))$$

The fact that S appears not to have changed after two flips in succession says that the function $M_{L'} \circ M_L$ shares a property with each of the mirror reflections:

$$(M_{L'} \circ M_L)(S) = S$$

In other words, the function $M_{L'} \circ M_L$ takes S onto itself; the function $M_{L'} \circ M_L$ leaves S apparently unchanged.

Can we say anything more about this new function? Our discussion up to now suggests a couple of lines of investigation.

Experiment 1

Take the square above with the two lines of mirror symmetry L and L'. The object of this experiment is to get a feeling for the function $F = M_{L'} \circ M_L$. You want to find out what F does to points. To help you do this, on a piece of paper draw a square and label the vertices with A, B, C, D. Draw in the lines L and L'.

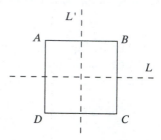

Then think of the function F as a "double flip." On a piece of tracing paper, trace a copy of the square—include the vertices in your tracing because you want to see where they go. First flip the traced copy about L. Record the labeling of the vertices now as compared to the original. Next flip the traced copy about L' (the original L'). Again, record the labeling of the vertices as compared to the original. What do you observe? Any conjectures?

Experiment 2

Another approach to finding out about $F = M_{L'} \circ M_L$ involves playing the comparison game again. This time let's compare F with an M_L itself since we now know a little bit about the M_L. We know that F shares some properties with M_L (and $M_{L'}$). For one thing it has the same domain and range. For another, $F(S) = S$ and $M_L(S) = S$. They both take S back onto itself! Is F like M_L in other ways? For example, does F preserve distances? What does F do to lines? Triangles? Circles? Does F reverse orientation? Could F be another $M_{L'}$ in disguise? If $X \in L$, then $M_L(X) = X$; i.e., X is a **fixed point** of M_L. What are the fixed points of F?

Experiment 3

A third approach to finding out about $F = M_{L'} \circ M_L$ is to use Geometer's Sketchpad. Set up the lines L and L' and an irregular polygon P. Have the software determine $M_L(P)$. Then have it determine $M_{L'}(M_L(P))$. Conclusions?

Let's try to get a feeling for $M_{L'} \circ M_L$ by following a point X through the two successive reflections. There are lots of choices for X. We'll choose X as in the diagram on the next page.

Assume first that line L and L' intersect in point C. Next, start with a point X. Suppose that the reflection of point X in line L is X' and that the reflection of point X' in line L' is X''. Thus,

$$M_{L'} \circ M_L(X) = M_{L'}(M_L(X)) = M_{L'}(X') = X''$$

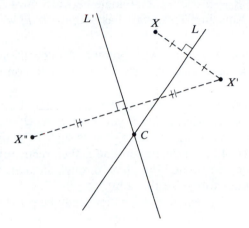

How are the points X, X', and X'' related? For one thing, the distances of the three points to C are equal: $d(X, C) = d(C, X') = d(C, X'')$. In other words, X, X' and X'' all lie on the same circle with center C. Interesting. Let's have a closer look.

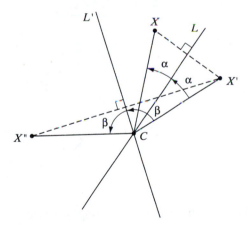

Let Y be the point of intersection of line XX' with L and Z the point of intersection of $X'X''$ with L'. Let $\alpha = \angle XCY$ and $\beta = \angle X''CZ$. Then it's easy to see that

$$\alpha = \angle XCY = \angle YCX' \text{ and } \beta = \angle X''CZ = \angle ZCX'$$

as in the diagram. Therefore, we conclude the following about the angle from X to X'' on the circle with center C.

$$\angle XCX'' = 2\beta - 2\alpha = 2(\beta - \alpha).$$

But $\beta - \alpha = \angle YCZ =$ angle between lines L and L' in the direction from L to L'. Amazing! This angle has nothing to do with the point X we started with!

Experiment 4   Start with some other points and follow them around as we did with X above. For example, take a point on L, then a point on L', then the point C. What happens?

It should seem obvious that, as a result of Experiment 3 and our "point chasing," no matter what point P you start with, $Q = F(P)$ will always be a point on the same circle—with center C—as P; angle PCQ will always be $2(\beta - \alpha)$. The combined effect of first reflecting in line L then reflecting the result in line L' is the same as if one had rotated the plane through an angle twice that from L to L' with center C! Let's formalize this a bit.

Let $R_{\phi,C}$ denote the function from the plane to the plane which for point X gives

$$R_{\phi,C}(X) = X'$$

where X and X' are on the same circle with center C and the angle from X to X' (measured in the counterclockwise direction) is ϕ. The function $R_{\phi,C}$ is called the **rotation through angle ϕ about center C.**

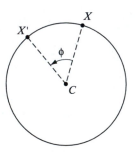

We have proved the following theorem.

Theorem 1.   If two lines L and L' intersect in point C and the counterclockwise angle from L to L' is ϕ, then

$$M_{L'} \circ M_L = R_{2\phi,C}$$

Thus if $M_{L'}(S) = S$ and $M_L(S) = S$, then also $R_{2\phi,C}(S) = S$ for a subset S.

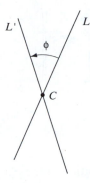

We'll come back to this theorem again. In the meantime, let's pause for an informal look at rotations and the notion of symmetry.

✳ *Rotational Symmetry*

Among the initial collection of shapes in the Problem for Right Now! you may have felt that some shapes had some kind of symmetry but not mirror symmetry. Here are some.

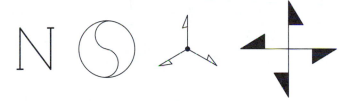

A discussion of these shapes might go something like this:

JERRY: The windmill has symmetry because you can turn it and it will still be the same.

CAROLE: What do you mean by "turn it"?

JERRY: Something like what we did with mirror symmetry: Trace the windmill on a piece of tracing paper. Pin the tracing paper to the original at the middle—where the lines cross—and turn the tracing paper counterclockwise through 90°. After you have done this, the traced windmill appears as if it has not changed; the traced shape fits exactly over the original.

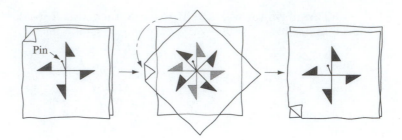

CAROLE: Neat!

This transformation of the plane—keeping one point fixed and turning the plane counterclockwise through a certain angle—is called a *rotation*. The fixed point C is called the *center* of the rotation and the angle α is called the *angle* of the rotation. A figure that remains unchanged after this transformation has taken place has *rotational symmetry with center C and angle α*. With this definition, the pinwheel has rotational symmetry with center at the point where the two bars cross and angle of 90°.

Problem 9

Trace a copy of each shape below and investigate its rotational symmetries in the same fashion as we used for the pinwheel above. For each rotational symmetry identify the center and angle.

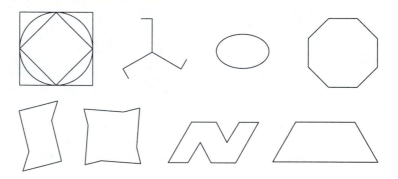

Now let's put the informal, transformational definition of rotation together with the rotation-as-function defined earlier. First, the 90° rotation of the pinwheel in the recent discussion corresponds to the function $R_{90°,C}$ defined on p. 236, in the sense that, if X is a point in the plane, then X is moved to the point $R_{90°,C}(X)$ by the transformation.

Second, Theorem 1 states that

$$M_{L'} \circ M_L = R_{2\phi,C}$$

where the angle from L to L' (counterclockwise) is ϕ and L and L' intersect at C. This is an equation of functions, meaning that

$$M_{L'} \circ M_L(X) = R_{2\phi, C}(X)$$

for every point X in the plane. It says that, even though the **actions** described by the left-hand side of the equation (follow a flip in L by a flip in L') and the right-hand side (rotate through angle 2ϕ about point C) are different, the results are the same. If you blinked while one was taking place, you wouldn't know which had happened by just looking at the result.

If shape S has bilateral symmetries about lines L and L', then $M_L(S) = S$ and $M_{L'}(S) = S$ and $M_{L'} \circ M_L(S) = M_{L'}(M_L(S)) = M_{L'}(S) = S$. Thus, also

$$R_{2\phi, C}(S) = S$$

This is another way of saying that if the plane is rotated through an angle 2ϕ about center C, then S is apparently unchanged. Thus, S has **rotational symmetry** $R_{2\phi, C}$.

For example, the square below has bilateral symmetries about line L and L'.

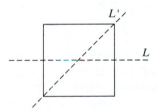

The angle from L to L' is $45°$. Thus, the square has rotational symmetry $R_{90°, C}$. Sure enough, the square has $90°$ rotational symmetry about its "center" C. (We knew that!)

For the pinwheel P with center C, $R_{90°, C}(P) = P$, even though P has no bilateral symmetries.

Problem 10

Use Geometer's Sketchpad to compare the two functions $R_{\alpha,C}$ and $M_{L'} \circ M_L$ where line L' and L intersect in C and the angle from L to L' (counterclockwise) is $\alpha/2$. For example, set up an irregular polygon and see what the two functions do to it. (Compare with problem 8.)

Problem 11

Every shape has rotational symmetry with center any point and angle 360°. (Why is this?) Such a symmetry of a shape is sometimes called the *trivial* symmetry; others are *nontrivial*. In this exercise, you want to pay attention to **all** the rotational and mirror symmetries of a shape. For example, in addition to the 90° rotation with center C, the pinwheel also has a symmetry of 180° rotation about C. It has even more rotational symmetries. (What are they?) List all the rotational and mirror symmetries of each shape below. Identify each rotational symmetry by its center and angle. Do you observe any relationships between the set of rotational symmetries and the set of mirror symmetries of a shape?

Here's another comment on Theorem 1. There, we started with a pair of bilateral symmetries M_L and $M_{L'}$ and arrived at their composition $M_{L'} \circ M_L = R_{2\phi,C}$. What if we **start** with a rotation $R_{\alpha,Q}$—the rotation of angle α about center Q? The theorem can be restated in the following manner.

Theorem 1′. Given point Q and angle α, let K, K' be any pair of lines intersecting in point Q and such that the counterclockwise angle from K to K' is equal to $\alpha/2$. Then $R_{\alpha,Q} = M_{K'} o M_K = M_{J'} o M_J$

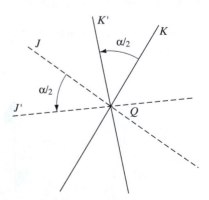

Consequently, $R_{\alpha,Q}$ can be written in many ways as a product of two reflections. This somewhat magical result will pop up later.

✳ *Kaleidoscopes*

Theorem 1 can also be used to describe and explain kaleidoscope phenomena. Recall that a simple kaleidoscope is formed by a pair of (real!) mirrors perpendicular to a plane $\mathscr{P}$. The two mirrors meet in a line in space and the overall design is created by a wedge W placed between the mirrors, the mirror reflections of the wedge, the mirror reflections of the mirror reflections, and so on. If L and L' denote the lines of intersection of the mirrors with the plane $\mathscr{P}$ (the plane of the paper), then $M_L(W)$ is the mirror reflection of W in mirror L and $M_{L'}(W)$ is the mirror reflection of W in mirror L'. Furthermore, $M_{L'}(M_L(W))$ is the mirror reflection of $M_L(W)$ in mirror L'. By the theorem, if the counterclockwise angle from L to L' is ϕ and L and L' intersect in C, then $M_{L'}(M_L(W)) = R_{2\phi,C}(W)$—what you get if you subject W to a rotation of twice ϕ about point C. You can see this in the diagram below, a bird's eye view of the mirrors looking down on plane $\mathscr{P}$.

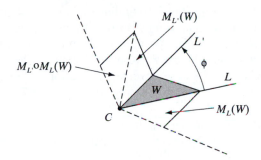

✳ *Successive Mirror Reflections in Two Parallel Lines*

We have just investigated a shape S having mirror symmetries in lines L and L'. We assumed that L and L' intersect in a point C. What happens if L and L' don't intersect but are parallel? The following experiments deal with this situation.

Experiment 5
In the case when L and L' intersected in point C we also had a shape S having mirror symmetries in the two lines. (Our example was a square.) This time—with L and L' parallel—we haven't produced a shape having mirror symmetries in the two lines. Does such a shape exist?

Experiment 6
Whether or not the shape in Experiment 5 exists, the function $F = M_{L'} \circ M_L$ **does** exist for L and L' parallel. The object of this experiment is to get a feeling for the function F. You want to find out what F does to points. To help you do

this, draw two parallel lines labeled L and L' on a piece of paper. Select five "representative" points on the plane more or less at random: one between the two lines, one on each of the lines, another to the right of both the lines, and a fifth to the left of both lines.

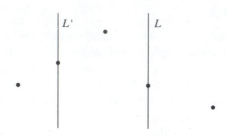

Then think of the function F as a "double flip." On a piece of tracing paper, trace the five points you selected above. First flip the traced copy about L. Next flip the traced copy about L' (the original L'). What do you observe? Any conjectures?

Experiment 7   Another approach to finding out about $F = M_{L'} \circ M_L$ when L and L' are parallel involves comparing F to those functions we already know something about: to M_L and to $R_{\alpha,C}$. We know that F has the same domain and range as the other functions. Does F preserve distances? What does F do to lines? Triangles? Circles? Does F reverse orientation? Could F be another $M_{L''}$ or an $R_{\alpha,C}$ in disguise? What are the fixed points of F?

Experiment 8   A third approach to finding out about $F = M_{L'} \circ M_L$ when L and L' are parallel is to use Geometer's Sketchpad. See Experiment 3 for ideas.

Let's see if we can find more information about the function $M_{L'} \circ M_L$. We know that it corresponds to following the flip of the plane about L by the flip about line L'. As we did when L and L' intersected in a point, let's follow a point X through two successive mirror reflections.

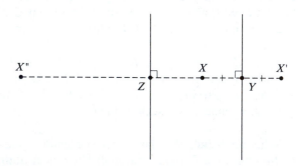

We denote the reflection of X in L by X' and the reflection of X' in L' by X''. In other words, $M_L(X) = X'$ and $M_{L'}(X') = X''$. Thus, also $M_{L'} \circ M_L(X) = M_{L'}(M_L(X)) = M_{L'}(X') = X''$. Denote also by Y the point where the perpendicular from X to line L intersects L; and denote by Z the point where the perpendicular from X' to L' intersects L'. Notice that X, Y, X', Z, X'' all lie on the same line! We know from the properties of mirrors that $d(X, Y) = d(Y, X')$ and $d(X', Z) = d(Z, X'')$ where $d(A, B) =$ distance between points A and B. Thus,

$$
\begin{aligned}
d(X, X'') &= d(X'', Z) + d(Z, X) \\
&= d(X', Z) + d(Z, X) \\
&= [d(Z, X) + d(X, Y) + d(Y, X')] + d(Z, X) \\
&= [d(Z, X) + d(X, Y) + d(Y, X)] + d(Z, X) \\
&= 2d(Z, X) + 2d(X, Y) \\
&= 2d(Z, Y)
\end{aligned}
$$

But $d(Z, Y)$ is just the distance between the lines L and L'. Once again, what happens to point X seems to be somewhat independent of X!

Problem for Right Now! For L and L' parallel, calculate

$$
M_{L'} \circ M_L(X)
$$

as we just did for several other choices of X. Here are the cases we haven't considered that you might want to try:

- X a point on line L
- X a point on line L'
- X a point to the right of L
- X a point to the left of L'

By examining the results of all cases above (and the results of Experiment 5), we arrive at the following conclusion.

Theorem 2. If L' and L are parallel lines and $X'' = M_{L'} \circ M_L(X)$, then

- X and X'' are on the same line perpendicular to L and L'
- $d(X'', X)$ equals twice the distance from L to L', measured along a line perpendicular to both
- On the line through X and X'', arrows from X to X'' and from L to L' point in the same direction.

If L and L' intersect, then we can describe a single, new function F such that $M_{L'} \circ M_L = F$. (In this case, F is a rotation.) Can the same thing be done when L and L' are parallel?

Let P and Q be two points in the plane and let $\mathbf{v}$ denote the vector with base P and tip Q.

Define the function $T_\mathbf{v}$ from the plane to the plane by setting $T_\mathbf{v}(X)$ equal to the fourth point of the parallelogram determined by one of its angles $\angle XPQ$.

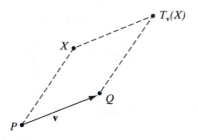

The function $T_\mathbf{v}$ is called **translation by v.**

If $T_\mathbf{v}(R) = S$ and $\mathbf{w}$ is the vector with base R and tip S, then $T_\mathbf{v} = T_\mathbf{w}$, i.e., the functions are the same and the vectors $\mathbf{v}$ and $\mathbf{w}$ are called **equivalent.**

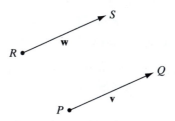

Using this definition, we can restate the above theorem as follows.

Theorem 3. Suppose L' and L are parallel lines, X is a point in the plane and $\mathbf{w}$ is the vector with base X and tip $X'' = M_{L'} \circ M_L(X)$. Then

$$M_{L'} \circ M_L = T_\mathbf{w}$$

The vector **w** points in a direction perpendicular to the two lines, from L to L'. The length of **w** is equal to twice the distance between the two lines. The function T_w in the theorem is independent of the choice of the point X and depends only on lines L and L'.

Problem 12 Use Geometer's Sketchpad to compare the two functions $M_{L'} \circ M_L$ and T_w of Theorem 3. See problems 8 and 10 for suggestions on how to do this.

Problem 13 Take a pair of **real** mirrors that are parallel and both perpendicular to a third plane. The silvered surfaces of the two mirrors should be facing each other. Place a design between the two silvered surfaces. Look at the design created by this original, its reflections, the reflections of the reflections, and so on. What do you see? How does the most recent theorem explain what you see? (Use the discussion in the "Kaleidoscopes" section of this chapter as a model.)

Problem 14 In Theorem 1', we pointed out that a given rotation $R_{\alpha,C}$ is equal to a "product" (i.e., a composition) $M_{K'} \circ M_K$ for **any** pair of lines K, K' satisfying certain properties related to α and C. What is an analogous result for a given translation T_u?

We began this discussion assuming there was a shape S of the plane having mirror symmetries in lines L and L'. In such a case, $M_L(S) = S$ and $M_{L'}(S) = S$. According to the recent theorem, if L and L' are parallel, then $T_w(S) = S$ also. In other words, S has *translational symmetry by vector* **v**. The following is an informal discussion of translational symmetry.

✳ *Translational Symmetry*

Consider a design suggested by the following picture.

⋯ $\rangle\rangle\rangle\rangle\rangle\rangle\rangle$ ⋯

The entire design extends indefinitely in both directions—to the left and to the right. The shape does not change when the entire plane undergoes the following transformation. Consider the arrow in the next diagram and the distance d between two like elements of the design.

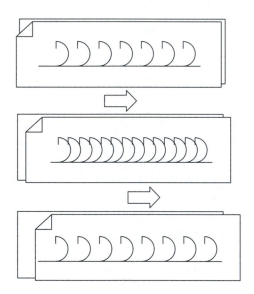

Trace a copy of the design on a separate sheet of paper. Then, starting with the tracing placed over the original design point for point, slide the tracing in the direction of the arrow for the distance d. The tracing should now fit exactly over the original.

The transformation of the plane consisting of a slide in the direction of the arrow through the distance corresponds to the function T_v where $\mathbf{v}$ is a vector of length d pointing in the same direction as the arrow.

Thus T_v takes the design back onto itself so that, consequently, the design has *translational symmetry by vector* **v**. Incidentally, the entire design has neither mirror symmetry nor rotational symmetry.

Problem 15   The design we have been working with has other translational symmetries in addition to the one described. What are they? (A different translation could have the same direction as the one described but a different distance. A different translation might also have a different direction but the same distance.) Make a tracing of the design on a separate sheet of paper to illustrate the translation described in the discussion above and to help you seek out these additional translations.

Here's another shape that has translational symmetry.

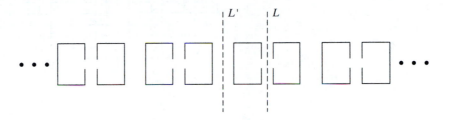

This shape also has mirror symmetries in many lines, two of which (L and L') are marked in the diagram above. This is a shape that we have been looking for, like one that you may already have found. In any case, $M_{L'} \circ M_L = T_w$, a translational symmetry of the shape.

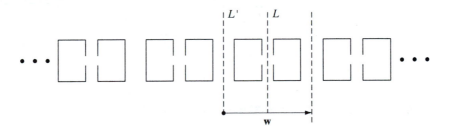

Problem 16   Describe the translational symmetries of each shape below. (It is possible that a shape has no translational symmetry.)

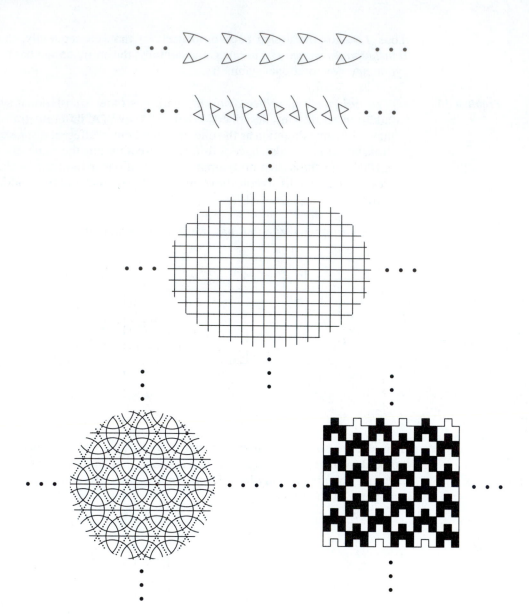

✳ *Products of Mirror Reflections*

Euclid's definition of a congruent triangle goes something like this. Triangle *ABC* is congruent to triangle *A'B'C'* if you can take a copy of the plane and move it so that triangle *ABC* fits exactly on top of triangle *A'B'C'*. For Euclid, the motion must be **rigid**—the plane must not be distorted, expanded, or shrunk. One way to ensure rigidity is to ensure that distances be preserved. We have been discussing three ways to move the plane: by mirror reflection, rotation, and translation. It seems pretty obvious that these motions preserve distances. We can see this from another point of view. Earlier we showed that each function M_L preserves distances. It is clear that the composition of two functions that preserve distance also preserves distance. (See Experiments 2 and 7.) As functions, rotations and translations are compositions of mirror reflections and so must also preserve distances.

For us we also think of a mirror reflection, a rotation, or translation as associated with a symmetry of a shape. We could use Euclid's definition of congruence to give us a general definition of the symmetry of shape S: Make a copy of the plane with a copy S' of shape S on it; then a rigid motion of the plane that fits S' exactly on top of S is a *symmetry* of S. The search for distance preserving functions is thus a search for functions which can be candidates for symmetries of shapes.

Are there other rigid motions that we haven't seen? One way to create more distance preserving functions is to compose (or "multiply") the distance preserving functions that we have:

$$T_{\mathbf{u}} \circ T_{\mathbf{v}}, \, T_{\mathbf{v}} \circ M_L, \, T_{\mathbf{u}} \circ R_{\alpha,C}, \ldots$$

But, since each factor of these products is itself an M_L or a composition of M_L's, their products must themselves be compositions of several M_L's:

If

$$T_{\mathbf{u}} = M_{L'} \circ M_L, \qquad T_{\mathbf{v}} = M_{K'} \circ M_K \quad \text{and} \quad R_{\alpha,C} = M_{J'} \circ M_J,$$

then

$$T_{\mathbf{u}} \circ T_{\mathbf{v}} = M_{L'} \circ M_L \circ M_{K'} \circ M_K,$$

$$T_{\mathbf{v}} \circ M_L = M_{K'} \circ M_K \circ M_L \quad \text{and}$$

$$T_{\mathbf{u}} \circ R_{\alpha,C} = M_{L'} \circ M_L \circ M_{J'} \circ M_J$$

All of these are "products" of mirror reflections. Anything else we could get from these functions by "multiplying" them would also be a product of mirror reflections. To make things simpler notationally, we will eliminate the "$\circ$" and denote the composition of two of these functions by juxtaposition. Thus, $T_{\mathbf{u}} \circ R_{\alpha,C}$ will be written $T_{\mathbf{u}}R_{\alpha,C}$. Let's have a closer look at these products in the following problems. We will denote by I the function that leaves everything unchanged: $I(X) = X$ for all points X in the plane. The function I is called the *identity function*.

Problem 17   What can you say about the function $M_L M_L$? (Is it equal to anything you have seen before?)

Problem 18   Suppose that L and L' are two lines. Do the functions M_L and $M_{L'}$ commute? In other words, is $M_L M_{L'} = M_{L'} M_L$ a true equation?

Problem 19 Earlier we discussed the fact that each M_L reverses orientation. What about $M_{L'}M_L$? What about $M_K M_{L'} M_L$? Can you generalize this?

Problem 20 Suppose that $\mathbf{P} = M_A M_B \ldots M_Y M_Z$ and $\mathbf{Q} = M_Z M_Y \ldots M_B M_A$. What can you say about the two functions $\mathbf{PQ}$ and $\mathbf{QP}$? (Problem 17 is a simple case of this.)

Problem 21 What can you say about the function $M_L R_{\alpha,C}$ when $C \in L$?

Problem 22 If f is a function whose domain and range is the plane, then a point X such that $f(X) = X$ is called a **fixed point** of f. What can you say about the fixed points of M_L? of $T_{\mathbf{u}}$? of $R_{\alpha,C}$?

Problem 23 Show that composition of functions is associative: that is, $F \circ (G \circ H) = (F \circ G) \circ H$. (Did you assume this fact in working on any of the previous problems?)

Problem 24 Show that if F is any function with domain and range equal to the plane, then $I \circ F = F \circ I = F$.

It seems a formidable task to investigate all possible products of mirror reflections in our search for more distance preserving functions (and, hence, symmetries). As we have seen in the preceding problems, some complicated-looking functions/ products are not really so complicated but are equal to familiar functions. Very simple examples of this are $R_{360°,C} = I$ and $T_{\mathbf{0}} = I$ (here, $\mathbf{0}$ is the zero-vector, the vector whose base and tip are the same). One technique we've used to investigate a function (a product?) is to investigate its set of fixed points. The set of fixed points of a mirror reflection is a line; the set of fixed points of a nontrivial rotation is a point; the set of fixed points of a nontrivial translation is the empty set. We might suspect that if the set of fixed points of a product is a line, then that product is in fact equal to a mirror reflection; or that if the set of fixed points is a point, then the product is a rotation; etc. Here is a theorem that enables us to tell a lot about a function based on very little information about its set of fixed points.

Theorem 4. Suppose that $\mathbf{P}$ is a product of mirror reflections and suppose that A, B, and C are three points in the plane not all on the same line. Suppose further that $\mathbf{P}(A) = A$, $\mathbf{P}(B) = B$ and $\mathbf{P}(C) = C$. (The points A, B, and C are fixed points of $\mathbf{P}$.) Then it follows that $\mathbf{P}(X) = X$ for all X in the plane. In other words, $\mathbf{P} = I$, the identity function.

Proof: We have already pointed out that $\mathbf{P}$ preserves distances, i.e., $d(X, Y) = d(\mathbf{P}(X), \mathbf{P}(Y))$ for all pairs points X, Y in the plane.

We will prove the theorem in stages. We will show first that $\mathbf{P}(A) = A$ and $\mathbf{P}(B) = B$ implies that $\mathbf{P}(D) = D$ for any point D on the line joining A and B. To do this, we know that

$$d(D, A) = d(\mathbf{P}(D), \mathbf{P}(A)) = d(\mathbf{P}(D), A)$$

In other words, D and $\mathbf{P}(D)$ lie on the same circle with center A. Call the radius of this circle r. Similarly, D and $\mathbf{P}(D)$ lie on the same circle with center B. Call the radius of this circle s.

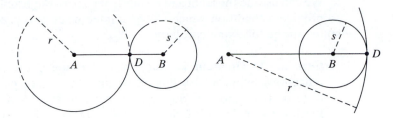

Two Cases

Consequently, D and $\mathbf{P}(D)$ are points of intersection of these two circles. But the two circles intersect in a single point. Thus, $D = \mathbf{P}(D)$.

Similarly, $\mathbf{P}(X) = X$ for any point X on the line joining A and C and for any point on the line joining B and C. Thus, we have that $\mathbf{P}(X) = X$ for any point X on the three lines extending the sides of the triangle ABC.

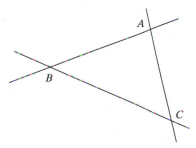

To complete the proof, we must show that $\mathbf{P}(Y) = Y$ for any point Y not on one of the three lines. Let Y be such a point. Then there is a line L through Y intersecting the three lines in at least two distinct points E and F. (Why is this so?)

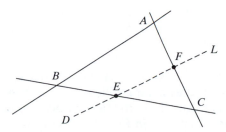

Since $\mathbf{P}(E) = E$ and $\mathbf{P}(F) = F$, the argument given above shows that any point on the line joining E and F is also fixed by $\mathbf{P}$. In particular $\mathbf{P}(Y) = Y$. This completes the proof of the theorem. ✳

We can use the theorem and results from one of the problems investigated earlier to provide a condition which will guarantee when two products are equal. This is stated in the following corollary.

Corollary to Theorem 4. Suppose that $\mathbf{P}$ and $\mathbf{Q}$ are both products of mirror reflections. Suppose also that A, B, and C are three points in the plane not all lying on the same line. Suppose further that $\mathbf{P}(A) = \mathbf{Q}(A)$, $\mathbf{P}(B) = \mathbf{Q}(B)$ and $\mathbf{P}(C) = \mathbf{Q}(C)$. Then it follows that $\mathbf{P} = \mathbf{Q}$; i.e., the two functions are equal.

Proof. Let $\mathbf{P}'$ be a product of mirror reflections such that $\mathbf{P}'\mathbf{P} = I$. The product $\mathbf{P}'$ exists as a consequence of problem 20 above.

If $\mathbf{P}(A) = \mathbf{Q}(A)$, then $A = I(A) = \mathbf{P}'\mathbf{P}(A) = \mathbf{P}'\mathbf{Q}(A)$. Similarly, $B = \mathbf{P}'\mathbf{Q}(B)$ and $C = \mathbf{P}'\mathbf{Q}(C)$. In other words,

$$A = \mathbf{P}'\mathbf{Q}(A), \; B = \mathbf{P}'\mathbf{Q}(B), \; C = \mathbf{P}'\mathbf{Q}(C)$$

Since $\mathbf{P}'$ and $\mathbf{Q}$ are products of mirror reflections, so is $\mathbf{P}'\mathbf{Q}$. The conditions of the theorem are satisfied. Thus, $\mathbf{P}'\mathbf{Q} = I$.

From $\mathbf{P}'\mathbf{Q} = I$, it follows that

$$\mathbf{P} = \mathbf{P}I = \mathbf{P}(\mathbf{P}'\mathbf{Q}) = (\mathbf{P}\mathbf{P}')\mathbf{Q} = I\mathbf{Q} = \mathbf{Q}$$

The associative and identity properties (problems 23 and 24) are used to justify this string of equations. This proves the corollary. ✳

The Corollary to Theorem 4 says that knowing what $\mathbf{P}$ does to the vertices of a triangle tells you a lot about $\mathbf{P}$. There is no other function—one that is a product of M_L's—which does the same as $\mathbf{P}$ does to those vertices. If we know exactly what $\mathbf{P}$ does to the vertices of a certain triangle, can we say anything more about $\mathbf{P}$? Is it a rotation? A translation? A mirror reflection? Which M_L's is $\mathbf{P}$ a product of? The following theorem goes a long way toward answering these questions.

Theorem 5. Suppose that triangle ABC is congruent to triangle $A'B'C'$. Then there is a product $\mathbf{P}$ of M_L's such that $\mathbf{P}(A) = A'$, $\mathbf{P}(B) = B'$, and $\mathbf{P}(C) = C'$. Moreover, the M_L's can be chosen so that there are at most three of them in the product.

Proof. Consider the congruent triangles ABC and $A'B'C'$ in Figure 1. Triangle ABC is drawn using solid lines (——); $A'B'C'$ is drawn using dotted lines (····).

Let L be the perpendicular bisector of the line segment BB'. If $B = B'$, this step is omitted. In Figure 2, line L as well as the reflection of triangle ABC in

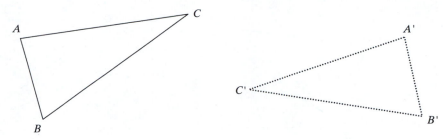

Figure 1

line L are drawn as dashed lines (----). The vertices of the reflected triangle are A'', B'', and C''. Of course, $M_L(A) = A''$, $M_L(B) = B''$, and $M_L(C) = C''$. Notice that $B' = B''$.

Let L' be the perpendicular bisector of the line segment joining C'' with C'. If $C'' = C'$, then this step is omitted. In Figure 3, line L' as well as the reflection of triangle $A''B''C''$ in line L' are drawn using dot-dash-dot lines (—·—·—). The vertices of the reflection of triangle $A''B''C''$ in L' are $\underline{A}$, $\underline{B}$, $\underline{C}$. So we have

$$M_{L'}(A'') = \underline{A},\ M_{L'}(B'') = \underline{B},\ M_{L'}(C'') = \underline{C}$$

Notice that $B' = B'' = \underline{B}$ and $C' = \underline{C}$. (Why so?)

Finally, let L'' be the perpendicular bisector of the line segment joining $\underline{A}$ with A'. If $\underline{A} = A'$, then this step is omitted. In Figure 3, line L'' is drawn as a dotted line (·····). Thus,

$$M_{L''}(\underline{A}) = A',\ M_{L''}(\underline{B}) = B',\ M_{L''}(\underline{C}) = C'$$

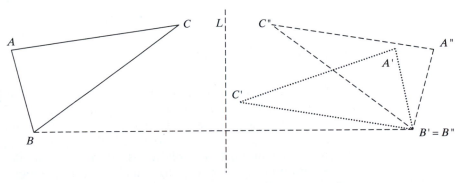

Figure 2

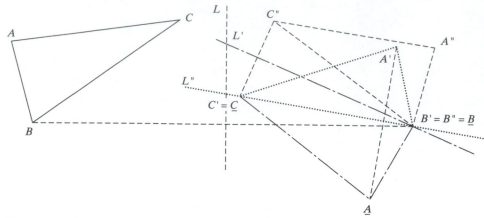

Figure 3

Thus the product $M_{L''}M_{L'}M_L$ takes A to A', B to B', and C to C'. This is what we wanted to show.

✳

Problem 25   On Geometer's Sketchpad set up two congruent triangles ABC and $A'B'C'$. Mimic the construction of Theorem 5: find the three lines L, L', and L'' and show that the product of the three reflections takes triangle ABC onto $A'B'C'$.

Theorem 5 says that given a pair of congruent triangles, there **is** a product of M_L's that takes one triangle to the other. The theorem even tells you what the M_L's in the product could be.

The following consequence of Theorems 4 and 5 will enable us to catalogue all products of M_L's.

Theorem 6.   Every product of M_L's is one of

$$I, \ M_L, \ M_LM_{L'}, \ M_LM_{L'}M_{L''}$$

for an appropriate choice of L, L', or L''.

Proof.   Let A, B, and C be three noncollinear points in the plane. Let **P** be a product of M_L's. Let $\mathbf{P}(A) = A'$, $\mathbf{P}(B) = B'$, and $\mathbf{P}(C) = C'$. Then triangle ABC is congruent to triangle $A'B'C'$. By Theorem 5 there is a function **Q** which is a "product" of 0, 1, 2, or 3 reflections such that $\mathbf{Q}(A) = A'$, $\mathbf{Q}(B) = B'$, and $\mathbf{Q}(C) = C'$. (A product of 0 reflections means **Q** is the identity; a product of one reflection means **Q** is equal to a single mirror reflection.) By the Corollary to Theorem 4, $\mathbf{P} = \mathbf{Q}$. Therefore, **P** is one of the types of functions listed.   ✳

Even though **P** may be written as the product of more than three reflections, the corollary says that **P** is equal (as a function) to the product of three or fewer reflections. For example, suppose that

$$\mathbf{P} = M_L M_K M_J M_H$$

for some lines L, K, J, and H. Then it's also true that either $\mathbf{P} = I$ or $\mathbf{P} = M_{L'}$ (for some line L') or $\mathbf{P} = M_{J'} M_{K'}$ (for some lines J' and K') or $\mathbf{P} = M_{L''} M_{K''} M_{J''}$ (for some lines L'', K'', and J'').

✳ *Where Are We? How Did We Get Here? Where Are We Going?*

We began this chapter asking "What is symmetry?" Since then we have pretty much followed our nose to get where we are. We had an instinctive feeling that we would recognize a symmetric shape when we saw one. Then we proceeded to pin these feelings down. First we defined mirror symmetry, associated it with a transformation, then a function from the plane to the plane. The world of functions turned out to be familiar and alien at the same time. Familiar, because we had worked with functions before. Alien, because these were not like functions we had worked with before: There were no graphs to look at, no tables of data to define them, and no algebraic formulas to give us insight. However, familiarity set us up to remember that one thing you can do with functions is compose them. This led us, on the one hand, to the definition of rotational symmetry. This confirmed an inkling we had felt earlier that some kind of symmetry was associated with rotation. On the other hand, composition of functions led us to the definition of translational symmetry, something new for most of us. We were led to accept it because we had created other ways of thinking about symmetry: Mirror and rotational symmetry corresponded to motions or transformation of the plane that left the shape unchanged; translational symmetry also corresponded to a motion that left the shape unchanged. Similarly, the functions corresponding to mirror and rotational symmetry take the shape in question back onto itself; the function corresponding to translational symmetry also takes the shape back onto itself. Moreover, all these functions are distance preserving.

Although we followed our nose in getting to where we are, I confess that I may have sneaked the notion of distance-preserving function into the discussion in an artificial manner. But I think it has turned out to be a unifying idea. It certainly helps make sense of Euclid's rigid motions.

As we "followed our nose," we used tactics that we found to be successful in earlier parts of the book. For example, we used analogy a lot: "This situation here is like that one there. In that one, such-and-such happens. Does it happen here?"

By following our nose we seem to have created an area of enquiry that is very rich: On the one hand, a lot of interesting things have turned up; on the other hand, there are a lot of unanswered questions. The simple, intuitive notion of symmetry has turned into a subject with a lot of "structure." Symmetries correspond to functions. Moreover, two of these symmetries, when multiplied, seem to give us another symmetry. Just like numbers. If we know the "factors," can we describe the "product"? We'll try to answer this question. Theorem 6 has left us both with a tool to answer it and also with another unanswered question: Can we describe all products of one-, two-, or three-mirror reflections? It looks like we'll be following our nose once again.

Here are some good warm-up problems for figuring out products:

Problem 26   What is an alternative description for the product $R_{\alpha,C}R_{\beta,C}$?

Problem 27   What is an alternative description for the product T_vT_w?

✳ *Products of Three Reflections*

All possible products of mirror reflections with an unlimited number of reflections in each product are really the same as all possible products with at most three reflections in each product. If the product has 0 factors, then $\mathbf{P} = I$; if one factor, then $\mathbf{P}$ is an M_L; if two factors, then $\mathbf{P}$ is either a rotation or a translation. Can we say something similar about $\mathbf{P}$ if it is the product of three reflections? Rotations and translations are surprising, alternate (simpler, perhaps?) descriptions of products of two reflections. Are there alternate (surprising? simpler?) descriptions of products of three reflections? Let's have a look.

First, let's see if a product of three reflections could be equal to one of the possibilities we already have: mirror reflection, rotation, or translation. One property of a mirror reflection we haven't taken advantage of is the fact that a mirror reflection reverses orientation. A consequence of this is that a product of two mirror reflections would reverse orientation twice and thus restore it. Thus, rotations and translations preserve orientation. What about a product of three reflections? The first two reflections in the product would preserve it, and the third would reverse orientation. The net effect of a product of three reflections would be to reverse orientation. This means that a product of three reflections could not be either a rotation or a translation. On the other hand, a product of three reflections **could** be equal to a mirror reflection. Can this happen? Yes! The product $M_LM_LM_{L'}$ is equal to $M_{L'}$! Does something like this always happen? Let's see.

Consider the function $\mathbf{P} = M_LM_KM_J$ where $L, K,$ and J are lines. To find alternate descriptions for $\mathbf{P}$, first note that the first part of the product is M_LM_K, which is either a rotation or a translation. Thus,

$$\mathbf{P} = R_{\alpha,C}M_J \quad \text{or} \quad \mathbf{P} = T_vM_J$$

Does this make things simpler? Maybe. It does give us two cases to examine.

Case 1: $\mathbf{P} = R_{\alpha,C}M_J$

In this case, there are two subcases: $C \in J$ and $C \notin J$. Let's look at the first of these, $C \in J$. Let K be a line through C so that the clockwise angle from J to K is $\alpha/2$.

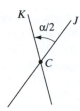

Thus, from a previous theorem, $R_{\alpha,C} = M_K M_J$. Consequently,

$$\mathbf{P} = R_{\alpha,C} M_J = M_K M_J M_J = M_K I = M_K$$

Theorem 7.   If $C \in J$, then $R_{\alpha,C} M_J = M_K$ where K is the line obtained by rotating J about C through a (clockwise) angle $\alpha/2$.

We will leave the other subcase of Case 1 for problem 31.

Case 2: $\mathbf{P} = T_v M_J$

We will deal with Case 2 in problems 28, 29, and 30.

Problem 28   What can you say about $T_v M_J$ in case $\mathbf{v}$ is perpendicular to J? [Hint: Above, in dealing with $R_{\alpha,C} M_J$ when $C \in J$, we wrote $R_{\alpha,C}$ as a product of reflections and we chose that product carefully (cf. a theorem about rotations proved earlier). Try something similar for T_v.]

Problem 29   What can you say about $T_v M_J$ when $\mathbf{v}$ and J are parallel? (Some helpful questions to ask: What does $T_v M_J$ do to orientation? What can you say about the fixed points of $T_v M_J$?)

Problem 30   What can you say about $T_v M_J$ in case $\mathbf{v}$ and J are **not** parallel and $\mathbf{v}$ is not perpendicular to J? [Hint: Suppose that $T_v = M_L M_K$. Then $T_v M_J$ is a product of three mirror reflections. The product $M_K M_J$ is a rotation. Replace K and J by another pair of lines that gives the same rotation but that may help you to get a handle on $T_v M_J$.]

Problem 31   What can you say about $R_{\alpha,C} M_L$ in case C is not a point on L? [Hint: See the hint for problem 30 with T_v replaced by $R_{\alpha,C}$.]

Problem 32   Find an alternative description of $R_{\alpha,C} R_{\beta,D}$, where C and D are distinct points. (Questions to ask: What does this product do to orientation? Does knowing the answer to the first question limit what the product can be? Some things to do: Try some special cases to see what can happen. Also, fiddle around with the lines that determine each rotation to see if you can't get something nice.)

Problem 33

Let F be a function whose domain is the plane and whose range is some subset of the plane. (The range may or may not be the whole plane.) Suppose also that F is distance preserving, i.e., $d(X, Y) = d(F(X), F(Y))$ for all pairs of points X and Y in the plane. (Mirror reflections, rotations, and translations are examples of distance-preserving functions but we do not assume that F is necessarily one of these.) Show that F must be a product of three or fewer reflections. (Hint: Show that Theorem 4 is true if **P** is replaced by any distance-preserving function. Then modify the corollary to Theorem 4 to get the result.)

Problem 34

Use problem 33 to show that any distance-preserving function has an inverse that is also distance-preserving. Function f' is an *inverse* to function f if and only if $ff' = f'f = I$.

Problem 35

Let S be a subset of the plane. Let G be the set of all distance-preserving functions f such that $f(S) = S$. Show that G has the following properties. (See problems 33 and 34.)

(a) If $f, g \in G$, then $fg \in G$ also.

(b) $I \in G$ and, of course, $If = fI = f$ for all $f \in G$.

(c) For all $f, g, h \in G, f(gh) = (fg)h$.

(d) For every $f \in G$, there exists $f' \in G$ such that $ff' = f'f = I$.

The set G described in problem 35 is called the **group of symmetries of** S. Let's see what the implications of problem 35 are in case the shape S is the rectangle below.

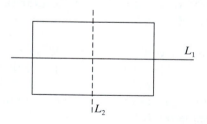

It seems reasonable that the set of symmetries of the rectangle is $G = \{I, R, M_1, M_2\}$ where R is rotation about the center C of the rectangle through 180° and M_1 and M_2 are mirror reflections in lines L_1 and L_2, respectively. We construct the "multiplication table" below to display all products of pairs of elements from the set G.

	I	M_1	M_2	R
I	I	M_1	M_2	R
M_1	M_1	I	R	M_2
M_2	M_2	R	I	M_1
R	R	M_2	M_1	I

The appearance of I in each row of the table verifies property (d) of the group of symmetries of the rectangle. In the next two problems you are asked to find the symmetries and construct the multiplication table for two additional shapes.

Problem 36   Find the group of symmetries of an equilateral triangle and construct its multiplication table.

Problem 37   Find the group of symmetries of a square and construct its multiplication table.

Problem 38   On the basis of the discussion in the text and the results of problems 28–31, you should have a complete catalog of distance preserving functions. What are they?

Problem 39   What can you say about $T_v R_{\alpha,C}$?

Problem 40   In a two-mirror kaleidoscope, the overall design is created by the mirror reflections in each of the two mirrors of an original shape S plus mirror reflections of mirror reflections and so on. Products of M_L's can be used to describe these multiple images of S as in the diagram below where the angle between the two mirrors is $\pi/6$. (Mirror reflections in mirrors 1 and 2 are abbreviated by M_1 and M_2, respectively.)

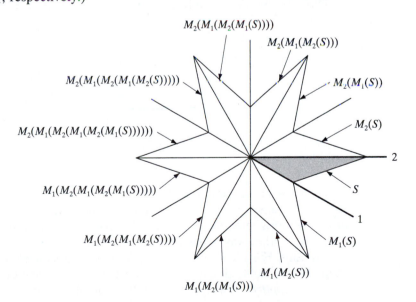

Using the results from Theorem 7 and problem 26, write simpler expressions for these multiple images. (See also the section on kaleidoscopes earlier in this chapter.)

✵ *Symmetries of Frieze Patterns and Wallpaper Designs*

This section consists primarily of experiments with shapes that have translational symmetry. Before starting on these, we should mention that the problems in the previous sections should have led you to the following two observations:

(1) The function $M_L T_v$ (L and v are parallel) reverses orientation and has no fixed points. Thus, it is neither a mirror reflection, nor a rotation, nor a translation; it is something new, called a **glide reflection** and denoted $G_{v,L}$. The following shape possesses the **glide reflectional symmetry** $G_{v,L}$.

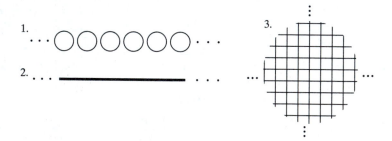

(2) Every product of mirror reflections (and, by problem 33, every distance preserving function) is equal to one of the following: the identity function, a mirror reflection, a rotation, a translation, or a glide reflection. Thus, every symmetry of a shape must also be one of these five types.

Three shapes that have translational symmetry are the following.

1. · · · ⃝⃝⃝⃝⃝ · · ·

2. · · · ▬▬▬▬▬ · · ·

3. (grid figure)

The first two shapes have translational symmetry in one direction only. The second shape has translational symmetry in more than one direction. (Two of these directions are horizontal and vertical.) The translational symmetries of the first two differ in another way. Consider the lengths of the vectors that correspond to the translations. For the shape on the left, there is a smallest nonzero length; for the shape on the right, all lengths are admissible. The shape on the left is an example of a *frieze pattern,* a shape that has translational symmetry in only one direction; moreover, there is a translation whose vector is of smallest nonzero length.

Experiment 9   Here is a bunch of frieze patterns. The object of this experiment is to describe the symmetries of each shape individually, then to see if there are properties of the symmetries that seem to prevail for all frieze patterns. If you can, work on this experiment in a group of three or four persons. The designs, from Kappraff, are patterns appearing on the pottery of San Ildefonso pueblo, New Mexico.

(1) Make a photocopy of the shapes. If you are working in a group, divvy up the shapes among the persons in your group. For each shape describe all the symmetries of the shape. List all the mirror reflectional symmetries and, on the shape itself, draw in the lines that go with them. List all the rotational symmetries; on the shape draw in the centers of these rotations. List all the translational symmetries. List all the glide reflectional symmetries.

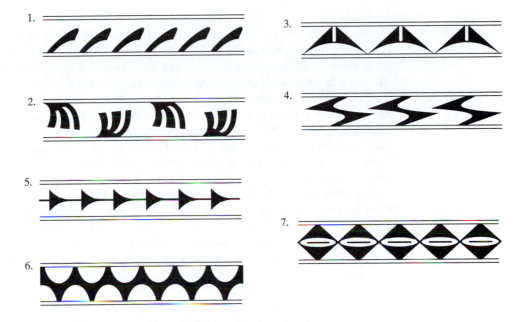

(2) After you have compiled the lists of symmetries for all the shapes, get your group together and compare the lists. Some questions to get you started: How do the mirror symmetries on the various lists compare? Is there anything in general you can say about them? How do the rotations compare? (What kinds of angles seem to be allowed in general? What can you say in general about the centers of rotations?) For each shape, how do you describe all the translations? How do you describe all the glide reflections?

A shape that has translational symmetry in more than one direction, such that for any direction there is a translation of smallest nonzero length, is called a *wallpaper design*. The following is an example of a shape that has translational symmetry in more than one direction but also has, in one of those directions, no nonzero translation of smallest length.

Experiment 10   On the next few pages are 17 wallpaper designs. The object of this experiment is to describe the symmetries of each design individually, then to see if there are properties of the symmetries that seem to prevail for all wallpaper designs. If there is a large class working on this, one way to tackle these designs is to divide the class up into groups of three or four persons. Choose one of the designs and have every group analyze it. Have the groups come back together as a class to compare results. Then divvy up the remaining designs among the groups, giving each group one or two designs to analyze. You might want to make enlarged copies of the designs so that when you draw on them you will be able to see what is going on.

For each design, describe all its symmetries. List the mirror symmetries and draw in the corresponding lines. List the rotational symmetries; on the shape draw in the centers of these rotations. List all the translational symmetries. List all the glide reflectional symmetries. If you are working on a design in a group, you might want to divvy up these tasks among the members of the group. For example, one person might want to work on the mirror reflections, another on the rotations, and so forth.

Note: Here is a device for displaying all the rotational symmetries for a single center in one fell swoop. For a given center C, look at all the angles of rotational symmetry with center C possessed by the design. There is usually a smallest non-zero angle. If this is so, then all the other angles of rotational symmetry with center C are multiples of this smallest angle. In particular, 360° is a multiple m of this smallest angle. In this case, C is called a *center of m-fold rotational symmetry*. For example, one design may have rotational symmetry with center C and angles 90°, 180°, 270°, and 360°. The smallest of these angles is 90°. Moreover, $360 = 4 \times 90$. So C is a center of fourfold symmetry for the design. In displaying the rotational symmetries of your design, all you need to do is mark the centers on the design and beside each center C place a numeral m, meaning that C is a center of m-fold rotational symmetry for the design.

The designs below are taken from Schattschneider, "The plane symmetry groups: . . . ."

You have observed many patterns in the symmetries of these wallpaper designs. No doubt you noticed something about their rotational symmetries: The only kinds of centers of rotational symmetry that appear are twofold, threefold, fourfold, and sixfold. Does this have to do with the particular designs we chose to analyze or is there something deeper happening? Let's find out.

First, notice that if a wallpaper design has one center of m-fold rotational symmetry, then it has infinitely many. (Why so?) Second, if $R_{\alpha,C}$ and $R_{\beta,D}$ are two rotations with C and D different, then $R_{\alpha,C}R_{\beta,D} = R_{\phi,E}$ where ϕ and E are as in the picture below (see Theorem 1′ and problem 32).

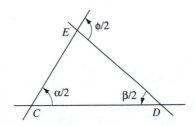

Third, if $R_{\alpha,C}$ and $R_{\beta,D}$ are rotational symmetries of the design, then so is their product $R_{\phi,E}$. In particular, E is also a center of rotational symmetry. Fourth, if P is a center of rotational symmetry for a wallpaper design and you consider all the distances of P to other centers, then there is a smallest positive distance. (Why so?)

Finally, let P be a center of p-fold rotational symmetry for a wallpaper design such that $p > 2$. Of all the centers of q-fold rotational symmetry such that $q > 2$ other than P itself, let Q be the closest one to P. So $R_{2\pi/p,P}$ and $R_{2\pi/q,Q}$ are rotational symmetries of the design; so is their product $R_{\delta,R}$; so is the latter's inverse $R_{2\pi/r,R}$. (Why so?) (We are not assuming that r is an integer.) Also R is a center of rotational symmetry for the design. Here's the picture:

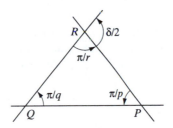

Now $\pi/p + \pi/q + \pi/r = \pi$. Thus, $1/p + 1/q + 1/r = 1$. (Does this look familiar?) There are two cases: (1) $p = q = 3$ and (2) $p > 2$, $q > 2$ and $p + q > 6$.

Case (1). If $p = q = 3$, then also $r = 3$. (Is there a wallpaper design in the collection above where this happens?)

Case (2). If $p, q > 2$ and $p + q > 6$, then $r < 3$ and consequently $r < q$. This means that the angle at R in triangle PQR is bigger than the angle at Q. This means that R is closer to P than Q is. Since Q was chosen to be closest to P (as a center of q-fold rotational symmetry with $q > 2$), this means that R is a center of twofold rotational symmetry. Thus, $r = 2$ and $1/p + 1/q = 1/2$. The solutions to this, as we have seen before, are $p = 3$, $q = 6$, and $p = q = 4$.

This analysis gives us the following theorem.

Theorem 8. Let C be a center of m-fold rotational symmetry of some wallpaper design. Then $m = 2, 3, 4,$ or 6.

Amazing!

✳ *Symmetries in Space*

In the final sections of this chapter, we want to discuss three-dimensional analogues to mirror reflections, symmetry, distance-preserving functions, and so on, in the plane. We will sketch only some of the possibilities in 3-space and leave most of the details for the problems. A large part of our discussion will focus on the symmetries of a regular tetrahedron and of a cube. These seemingly simple figures are rich in symmetry—there will be surprises. Working with these shapes will suggest questions for further study in three-dimensional symmetry and point out features of three-dimensional symmetry that were not predicted by what goes on in two dimensions. It will also give us new insight into the shapes themselves.

✳ *Mirror Reflections in Space*

As in the plane, we can define a function from 3-space to 3-space that captures something of what a real mirror does. Let $\mathscr{P}$ be the plane of a mirror. If X is a point in space, then the **mirror image** X' in the mirror can be obtained in the following way. Drop a perpendicular from X to $\mathscr{P}$. Let this intersect $\mathscr{P}$ in point Y. Then extend this line the same distance on the other side of $\mathscr{P}$ to obtain point X'. Thus, $d(X, Y) = d(X', Y)$. We define the corresponding function $M_{\mathscr{P}}$ by

$$M_{\mathscr{P}}(X) = X'$$

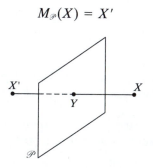

Problem 41   Show that $M_{\mathscr{P}}$ is a distance preserving function.

Problem 42   Suppose that F is a distance-preserving function from space to space. Let L be a line, $\mathscr{P}$ a plane, C a circle, and S a sphere in space. Show that $F(L)$ is a line, $F(\mathscr{P})$ a plane, $F(C)$ a circle and $F(S)$ a sphere.

✳ *Rotations in Space*

What should a rotation of space be? In the plane, a rotation is the product of two-mirror reflections whose lines are not parallel. In space, consider the product of two-mirror reflections whose planes aren't parallel. But, two nonparallel planes can't intersect in just a point; they must intersect in a line. Thus, take $M_{\mathscr{P}}$ and $M_{\mathscr{P}'}$ such that planes $\mathscr{P}$ and $\mathscr{P}'$ intersect in line L. Let's see what the function $M_{\mathscr{P}}M_{\mathscr{P}'}$ does to a point X in space. First extend a perpendicular from X to $\mathscr{P}'$ an equal distance on the other side of $\mathscr{P}'$ to a point X'. Any plane containing X and X' is perpendicular to $\mathscr{P}'$. Then extend a perpendicular from X' to $\mathscr{P}$ an equal distance on the other side of $\mathscr{P}$ to a point X''. Any plane containing X' and X'' is perpendicular to $\mathscr{P}$. Thus, points X, X', and X'' all lie in a plane perpendicular to *both* $\mathscr{P}$ and $\mathscr{P}'$. This is the unique plane $\mathscr{P}''$ through X perpendicular to line L. All the "action" for X "takes place" in plane $\mathscr{P}''$. Let's have a closer look at this action. First, let C be the intersection of line L with plane $\mathscr{P}''$.

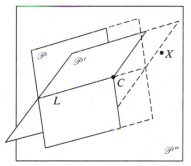 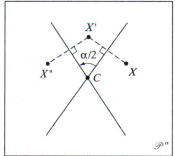

Looking "head on" at
plane $\mathscr{P}''$ through line L

As can be seen from the diagram, X'' is the result of rotating X in plane $\mathscr{P}''$ about point C through twice the dihedral angle from $\mathscr{P}'$ to $\mathscr{P}$. The effect of this on all of space is what is called a **rotation about line L through angle α**. (In this case α is twice the dihedral angle from $\mathscr{P}'$ to $\mathscr{P}$.) The corresponding function is denoted $R_{\alpha,L}$. The line L is called the **axis** of the rotation. We have the following theorem.

Theorem 9.   Let $\mathscr{P}$ and $\mathscr{P}'$ be two planes intersecting in line L. Let the dihedral angle from $\mathscr{P}'$ to $\mathscr{P}$ be β. Then

$$M_{\mathscr{P}}M_{\mathscr{P}'} = R_{2\beta,L}$$

Problem 43   Given a rotation $R_{\alpha,C}$ in the plane, it's true that $R_{\alpha,C} = M_{L'}M_L$ for any pair of lines that intersect in C and such that the angle from L to L' is $\alpha/2$. What is the analogue to this fact in space?

Problem 44   If $\mathscr{Q}$ is a plane containing the line L, what is an alternate description of $R_{\alpha,L}M_{\mathscr{Q}}$?

Problem 45   Suppose that planes $\mathscr{P}$ and $\mathscr{P}'$ are parallel. What can you say about the product $M_{\mathscr{P}}M_{\mathscr{P}'}$?

Problem 46   What is an alternate description of $R_{\alpha,L}R_{\beta,L}$?

✷ *Symmetries of a Regular Tetrahedron, Part I*

As in the plane, a subset S of space has **mirror symmetry in plane** $\mathscr{P}$ means that $M_{\mathscr{P}}(S) = S$. The subset S has **rotational symmetry about line L through angle α** means that $R_{\alpha,L}(S) = S$. Let's look for the rotational and mirror symmetries of a regular tetrahedron.

Take a regular tetrahedron with vertices A, B, C, and D and look at it "head on" through vertex D.

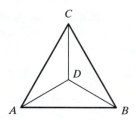

Consider the line L_D through D perpendicular to and passing through the center of face ABC. Line L_D is an axis of rotational symmetry for the tetrahedron, through angles $2\pi/3$ and $4\pi/3$. Thus, $R_{2\pi/3,L_D}$ and $R_{4\pi/3,L_D}$ are symmetries of the tetrahedron. Define lines L_A, L_B, and L_C similarly. These are also axes of rotational symmetry, also through angles $2\pi/3$ and $4\pi/3$. This gives us a total of eight rotational symmetries. Add in identity I and we get nine in all.

For planes of mirror symmetry consider the plane $\mathscr{P}_{CD}$ determined by C, D and the midpoint of AB. This plane is perpendicular to face ABC. It is also a plane of mirror symmetry for the tetrahedron. So are the planes $\mathscr{P}_{AB}$, $\mathscr{P}_{CB}$, $\mathscr{P}_{AD}$, $\mathscr{P}_{AC}$, and $\mathscr{P}_{BD}$, defined similarly. Thus, the tetrahedron has six mirror symmetries.

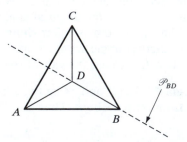

Next, "stand" the tetrahedron on edge *CD* and look "through" edge *AB*.

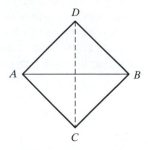

The line $L_{AB,CD}$ passing through the midpoints of *AB* and *CD* is an axis of rotational symmetry through angle π. Similarly defined lines $L_{AC,DB}$ and $L_{AD,CB}$ are also axes of rotational symmetry both with angles π. This adds three more rotational symmetries for a total of 12.

Notice that planes $\mathscr{P}_{AB}$ and $\mathscr{P}_{CD}$ are perpendicular and that

$$M_{\mathscr{P}_{AB}} M_{\mathscr{P}_{CD}} = R_{\pi, L_{AB,CD}}$$

So far we have $12 + 6 = 18$ symmetries. Are there more? In the plane we know that the product of two symmetries is also a symmetry. This is true in space as well. Maybe products of known symmetries of the tetrahedron will give us something new. Let's take a side excursion to look at some products independent of the tetrahedron.

✸ *Product of a Mirror Reflection with a Rotation*

Suppose that $\mathscr{P}$ is a plane, *L* is a line, and α is an angle. Can we say anything about the product

$$R_{\alpha,L} M_{\mathscr{P}}$$

Before analyzing this in detail, we should discuss the sort of thing we might expect. Again, our experience with mirror reflections and rotations in the plane is our guide. Recall that a mirror reflection in the plane reverses orientation. The same happens with a mirror reflection in space. In space, an orientation is given by a 3-dimensional coordinate system.

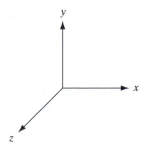

This is a right-handed coordinate system: x-axis (right thumb), y-axis (right forefinger), and z-axis (right middle finger). Under a mirror reflection this becomes a left-handed coordinate system: x-axis (left thumb), y-axis (left forefinger), and z-axis (left middle finger). Moreover, a left-handed coordinate system becomes a right-handed system under a mirror reflection. A normal clock with an opaque back under a mirror reflection will turn into a clock that reads counterclockwise.

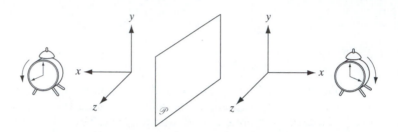

A rotation being the product of two reflections will preserve orientation. Finally, the product of a rotation with a mirror reflection will reverse orientation. This allows the possibility that the product of a rotation with a mirror reflection is another mirror reflection. If that happens, then the product would have a whole plane of fixed points. It's time to have a close look at this product. Let's do it case by case.

Case 1. L is a subset of $\mathscr{P}$. Problem 44 deals with this.

Case 2. (Very special case) L is perpendicular to $\mathscr{P}$ and intersects it in point p; the angle of the rotation is π.

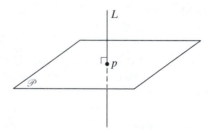

It's easy to see, by following a typical point around, that the only fixed point of $R_{\pi,L}M_{\mathscr{P}}$ is p. This is something new! In fact, if $p = (0, 0, 0)$, the origin, and $X = (x, y, z)$, then

$$R_{\pi,L}M_{\mathscr{P}}(x, y, z) = (-x, -y, -z)$$

In this case the function $R_{\pi,L}M_{\mathscr{P}}$ is called **inversion through point** p and we write

$$R_{\pi,L}M_{\mathscr{P}} = I_p$$

Case 3. L and $\mathscr{P}$ are perpendicular; the angle α is arbitrary.

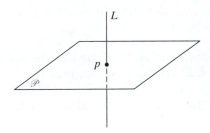

Again, by following a typical point around, we see that the only fixed point is p. Still, something new. However, it is not difficult to see that $R_{\alpha,L}M_{\mathscr{P}} = I_p R_{\alpha-\pi,L}$. This function is called a **rotary inversion through point p, about line L, through angle** $\alpha - \pi$.

Case 4. (General case) L and $\mathscr{P}$ intersect in point p; angle α arbitrary.

Let $\mathscr{P}'$ and $\mathscr{P}''$ be two planes intersecting in line L such that the dihedral angle from $\mathscr{P}'$ to $\mathscr{P}''$ is $\alpha/2$ and such that $\mathscr{P}'$ is perpendicular to $\mathscr{P}$ (see problem 31). Then $R_{\alpha,L}M_{\mathscr{P}} = M_{\mathscr{P}''}M_{\mathscr{P}'}M_{\mathscr{P}}$. Now $M_{\mathscr{P}'}M_{\mathscr{P}}$ is a rotation through angle π. Again, by problem 43, $M_{\mathscr{P}'}M_{\mathscr{P}} = M_{\mathscr{Q}'}M_{\mathscr{Q}}$ where $\mathscr{Q}'$ and $\mathscr{Q}$ intersect in the same line $\mathscr{P}'$ and $\mathscr{P}$ do, $\mathscr{Q}$ is perpendicular to both $\mathscr{P}''$ and $\mathscr{Q}'$. Thus

$$R_{\alpha,L}M_{\mathscr{P}} = M_{\mathscr{P}''}M_{\mathscr{P}'}M_{\mathscr{P}} = M_{\mathscr{P}''}M_{\mathscr{Q}'}M_{\mathscr{Q}}$$

The function $M_{\mathscr{P}''}M_{\mathscr{Q}'}$ is a rotation with axis perpendicular to plane $\mathscr{Q}$. Thus $R_{\alpha,L}M_{\mathscr{P}} = M_{\mathscr{P}''}M_{\mathscr{Q}'}M_{\mathscr{Q}}$ falls under Case 3 and is a rotary inversion.

Problem 47 Given α, L, and $\mathscr{P}$ in Case 4, how are the line and the angle of the resulting rotary inversion determined?

Problem 48 Suppose that line L is parallel to plane $\mathscr{P}$. What is an alternate description of $R_{\alpha,L}M_{\mathscr{P}}$?

Problem 49 Let F be a distance preserving function from space to space. Let A, B, C, D be the vertices of a tetrahedron. Suppose that $F(A) = A$, $F(B) = B$, $F(C) = C$ and $F(D) = D$. Then $F = I$, the identity function. Show this!

Problem 50 Suppose that $ABCD$ and $A'B'C'D'$ are two congruent tetrahedra. Then there is a function F which is the product of four or fewer mirror reflections such that $F(A) = A'$, $F(B) = B'$, $F(C) = C'$, $F(D) = D'$. Show this!

Problem 51 Suppose that F is a function from space to space which is also the product of several mirror reflections. Show that F has an inverse. What **is** the inverse?

Problem 52   Suppose that F is a distance-preserving function from space to space. Show that F is equal to a function which is the product of four or fewer mirror reflections. (Hint: See the proof of the analogous result in the plane.)

Problem 53   Suppose that A, B, C, and D are the vertices of a tetrahedron. Suppose that F and G are two distance-preserving functions from space to space such that $F(A) = G(A)$, $F(B) = G(B)$, $F(C) = G(C)$, and $F(D) = G(D)$. Show that $F = G$. (Hint: See the proof of the analogous result in the plane.)

✳ *Symmetries of the Regular Tetrahedron, Part II*

As a consequence of problem 53, we have the following theorem.

Theorem 10.   The regular tetrahedron has at most 24 symmetries.

Proof.   We are assuming, of course, that a symmetry of the regular tetrahedron T is a distance preserving function F such that $F(T) = T$.

Suppose that the vertices of T are A, B, C, and D. If F is a symmetry of T, then the points $F(A)$, $F(B)$, $F(C)$, and $F(D)$ constitute some rearrangement of the vertices of T. In other words, $\{F(A), F(B), F(C), F(D)\}$ is a permutation of $\{A, B, C, D\}$. Since there are $n!$ permutations of n things, there are $4! = 24$ permutations of $\{A, B, C, D\}$. By problem 53, at most one distance preserving function is associated with each permutation.   ✳

We have found that the regular tetrahedron has 12 rotational symmetries (including I) and six mirror symmetries, for a total of 18 symmetries. Suppose we label the 12 rotational symmetries by $R_1, \ldots, R_{12}$ and let M denote one of T's mirror symmetries. Then consider the following functions obtained by multiplying each of the rotations by M:

$$MR_1, \ldots, MR_{12}$$

What can we say about these functions? All are symmetries of T. None is a rotation. (Why?) So there is no overlap in the two lists $R_1, \ldots, R_{12}$, and $MR_1, \ldots, MR_{12}$. Finally, the functions on the list $MR_1, \ldots, MR_{12}$ are all distinct. (Proof: Suppose that

$$MR_i = MR_j$$

for some i and j. "Multiply" both sides of the latter equation by M:

$$MMR_i = MMR_j$$

or, since $MM = I$ (why?),

$$IR_i = IR_j$$

or

$$R_i = R_j$$

In other words, $i = j$.)

Thus, the 24 functions $R_1, \ldots, R_{12}, MR_1, \ldots, MR_{12}$ must all be distinct symmetries of T. Since there can be at most 24, this must be a complete list.

What can we say about this list? We know that $R_1, \ldots, R_{12}$ preserve orientation and that $MR_1, \ldots, MR_{12}$ reverse it. Thus, the six mirror symmetries, which also reverse orientation, must be included in $MR_1, \ldots, MR_{12}$. That leaves six of the $MR_1, \ldots, MR_{12}$ to be rotary inversions.

Here is an example of a rotary inversion which is a symmetry of T. Look at T "edge on" once again.

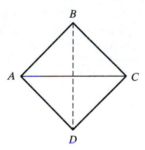

Let L be the line joining the midpoints of AC and BD. Rotate T about axis L through angle $\pi/2$ to get the following.

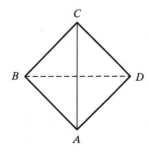

Then invert the result through the center p of T to get

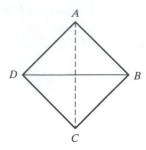

Thus, $I_p R_{\pi/2,L}$ is a symmetry of T. So also is $I_p R_{3\pi/2,L}$. There are three distinct lines joining midpoints of opposite edges of T. This yields six rotary inversions which are symmetries of T. Our list of symmetries of T is complete!

Problem 54   Show that $I_p R_{\pi/2,L} I_p R_{\pi/2,L} = R_{\pi,L}$. Give three reasons why $I_p R_{\pi,L}$ is not a symmetry of T.

✳ *Symmetries of the Cube, Part I*

As we did with the regular tetrahedron, let's list some of the "obvious" symmetries of the cube. Let's start with mirror symmetries.

Consider a plane parallel to and halfway between two opposite faces. This is a plane of mirror symmetry. The cube has six faces, thus three pairs of opposite faces. So there are three mirror symmetries of this kind.

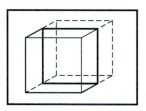

Next consider a plane passing through opposite edges. This is a plane of mirror symmetry. The cube has 12 edges and so six pairs of opposite edges. Thus, there are six mirror symmetries of this kind.

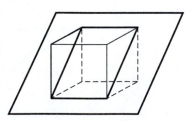

So nine mirror symmetries in all—so far. Let's turn to rotational symmetries.

Consider a line joining midpoints of opposite faces. This is an axis of rotational symmetry through angles $\pi/2$, π, and $3\pi/2$. The cube has six faces, so three pairs of opposite faces. Thus, there are nine rotational symmetries of this kind.

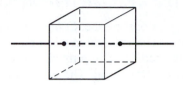

Next consider a line joining midpoints of opposite edges. This is an axis of rotational symmetry through angle π. The cube has 12 edges and therefore six pairs of opposite edges. Thus, there are six rotations of this kind.

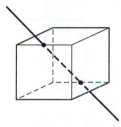

Now for a surprise. Consider a line joining opposite vertices. This is an axis of rotational symmetry through $\pi/3$ and $2\pi/3$. The cube has eight vertices and four pairs of opposite vertices. Thus, there are eight rotations of this kind. (The surprise is that a cube—old "square face"—shares symmetries with a regular tetrahedron.)

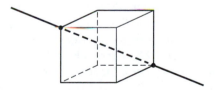

That makes $9 + 6 + 8 = 23$ rotational symmetries. Add the identity and that makes 24 rotational symmetries in all.

We have nine mirror symmetries and 24 rotational symmetries—33 in all, so far.

Before going further, let's try analyzing the symmetries of the cube as we did the symmetries of the regular tetrahedron. Denote the 24 rotational symmetries of the cube by $R_1, R_2, \ldots, R_{24}$. Multiply all of these by M, one the cube's mirror symmetries, to obtain $MR_1, MR_2, \ldots, MR_{24}$. By arguments similar to those for the tetrahedron, the functions $R_1, R_2, \ldots, R_{24}, MR_1, MR_2, \ldots, MR_{24}$ are distinct symmetries of the cube. This says two things. First, that the cube has 48 or more symmetries. Second, that our list of 33 symmetries is incomplete. What are the symmetries that we have missed? It turns out that the symmetries of a regular tetrahedron will help us complete our discussion of the symmetries of the cube.

✳ Symmetries of the Regular Tetrahedron and the Cube, Concluded

Select a vertex of the cube. Connect that vertex to three others along face diagonals of the cube. The original vertex and the three others also form the vertices of a regular tetrahedron. The remaining four vertices of the cube also form the vertices of a regular tetrahedron. The two tetrahedra are congruent.

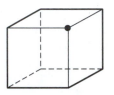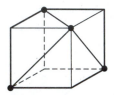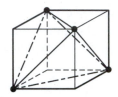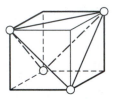

Since a symmetry of the cube will take the cube's vertices back onto themselves, it will also either switch the two tetrahedra or it will be a symmetry of both tetrahedra, that is, take each tetrahedron back onto itself.

One symmetry of the cube switches the two tetrahedra nicely. It's not on our list. It's inversion about the center of the cube. It's clearly a symmetry of the cube and it must switch the two tetrahedra (if for no other reason than that it can't be a symmetry of either tetrahedron).

How many symmetries of the cube are also symmetries of one (and hence the other also) tetrahedron? Put another way, how many symmetries of one of the tetrahedra are also symmetries of the cube? It's not difficult to see that all the rotational and mirror symmetries of one of the tetrahedra are also symmetries of the cube. That's all that's necessary in order for all the symmetries of the tetrahedron to be symmetries of the cube. (The remaining symmetries of the tetrahedron are, after all, products of its rotational and mirror symmetries.)

Thus, exactly 24 of the cube's symmetries also take both inscribed tetrahedra back onto themselves. Denote these symmetries by $S_1, \ldots, S_{24}$. Multiply each of these by I_p to get $I_pS_1, \ldots, I_pS_{24}$. The latter are all symmetries of the cube that switch the two tetrahedra. By an argument similar to one we have seen, the symmetries $I_pS_1,$ $\ldots, I_pS_{24}$ are all distinct. Furthermore, if Q is a symmetry of the cube which switches the two tetrahedra, then I_pQ is a symmetry that doesn't switch the tetrahedra. Thus,

$$I_pQ = S_i$$

for some i. But then

$$I_pI_pQ = I_pS_i$$

or, since $I_p I_p = I$,

$$Q = I_p S_i$$

Consequently, every symmetry of the cube which switches tetrahedra is one of $I_p S_1$, . . . , $I_p S_{24}$. Thus all symmetries of the cube are accounted for in the set

$$\{S_1, \ldots, S_{24}, I_p S_1, \ldots, I_p S_{24}\}.$$

There are exactly 48 elements in this set. So the cube has exactly 48 symmetries.

Another way to list the symmetries of the cube is to first list the rotations R_1, . . . , R_{24} and follow them by $I_p R_1$, . . . , $I_p R_{24}$. The symmetries are easier to figure out this way. The answer to the following problem may help.

Problem 55   Suppose $p \in L$. What conditions on α make $I_p R_{\alpha, L}$ a mirror reflection? (What's the plane of the mirror reflection?)

Problem 56   Which symmetries (of the cube) in the list R_1, . . . , R_{24}, $I_p R_1$, . . . , $I_p R_{24}$ switch the tetrahedra and which don't?

✳ *A Three-Dimensional Kaleidoscope*

The n lines of mirror symmetry of a regular n-gon in the plane dissect the regular n-gon into $2n$ congruent pieces, each one a right triangle. The number $2n$ of pieces is also equal to the number of symmetries possessed by the regular n-gon. If you place mirrors along the sides of one of these pieces and meeting at the center of the polygon, then the piece of the polygon and its mirror images will "re-create" the whole regular n-gon as the total design in the kaleidoscope.

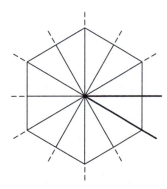

Can we mimic this with the cube? The following diagram shows all the planes of mirror symmetries of the cube.

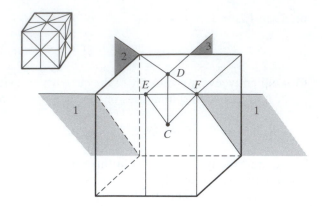

These planes dissect the cube into 48 "congruent" pieces, each one a triangular pyramid having one vertex at the center of the cube. If you place mirrors along the sides of one of these pyramids so that the mirrors meet at the center of the cube, what do you get? The pyramid and its mirror images will "re-create" the whole cube as the total design in the kaleidoscope! This takes seeing to believe and appreciate. You need to make (or have a friend make) such a 3-D kaleidoscope. In the meantime, let's have a look at the three planes that make up this magic device. Note that the three planes of mirror symmetry meet at the center of the cube. The dihedral angle between planes 2 and 3 is $\pi/4$. (Their product is a rotation of angle $\pi/2$ with axis DC through the center D of the face and the center C of the cube.) The dihedral angle between planes 1 and 3 is $\pi/2$. (Their product is a rotation of angle π with axis EC, where E is the midpoint of an edge.) Finally, the dihedral angle between planes 1 and 2 is $\pi/3$. (Their product is a rotation of angle $2\pi/3$ with axis FC, where F is a vertex of the cube.) So there are three planes meeting in a point with dihedral angles $\pi/2$, $\pi/3$, and $\pi/4$. This is one of the admissible 3-mirrors-meeting-in-a-point kaleidoscopes that we met in the kaleidoscope chapter! Incredible! (What do you suppose the other admissible kaleidoscopes correspond to?)

✳ *Full Circle*

This chapter is full of functions, utilizing a point of view new to the book and somewhat unexpected in geometry (to most readers anyway). Functions were used to formalize mirror symmetry and to put a new slant on how a flat mirror behaves. Knowledge of functions, part of any mathematician's (young or old) baggage, gave us a framework with which to investigate these new functions ("What happens when you compose them?"). What emerged, however, was a new approach to geometry with an unexpected algebraic flavor. This new approach enhanced and deepened our intuitive notions of symmetry. It gave us a new way to look at kaleidoscopes. It also gave us fresh insight into the old world of functions: We became interested in their fixed points, their distance-preserving features, and whether or not they preserved orientation. We were able to prove several theorems regarding these functions. A big one states that knowing what a distance-preserving function does to the vertices of a triangle uniquely determines the function. Another big theorem cataloged all possible distance-preserving functions in the plane. Our success with this new approach led us on the one hand to investigate the symmetries of frieze and wallpaper patterns and on the other to generalize what had been done in the plane to 3-space. Both of these latter studies were not intended to be complete but rather to suggest the richness of this new approach to geometry and open up possibilities for further exploration.

✳ *Notes*

Symmetry has been around a long time, in the form of symmetrical shapes in nature and in human artifacts.

Reproduced by permission from *Art of a Vanished Race: The Mimbres Classic Black-On-White* (Dillon Tyler Publishers, 1975)

Leonardo da Vinci may have been the first to study symmetry systematically. He classified architectural floor plans according to the symmetries they possess. For example, there is the class of all the floor plans having the same symmetries as the square.

Prototypes of the other symmetry classes, ones with a finite number of symmetries, are the regular polygons and the regular polygons "with flags."

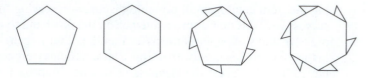

Serious mathematical study of symmetry took off in the nineteenth century. The mathematical theory of rigid motions (distance preserving functions, and "products" of same) was developed by mathematicians Chasles, Rodrigues, Cayley, Sylvester, Hamilton, and Donkin. The concept of the group of symmetries of a shape was created by Möbius and Hessel. Like Leonardo da Vinci for the plane, Hessel in 1830 enumerated the classes of shapes in space that have a finite number of symmetries. Independently, in 1856 William Rowan Hamilton found all the symmetries of the regular polyhedra (Fejes Tóth, 1964).

In his *Erlanger Programm* of 1872, the German mathematician Felix Klein proposed that the proper study of any geometry is the group of functions that preserves its structure. For Euclidean plane geometry, this is the set of distance preserving functions. The foundations for the theory of groups had been laid in the early part of the nineteenth century by Lagrange, Ruffini, Abel, Galois, and Cauchy (Fejes Tóth, 1964).

The first direct mathematical treatment of wallpaper designs was given by the Russian mathematician Fedorov in 1891 as part of his larger study of crystallography. As Leonardo da Vinci and Hessel before him, he classified wallpaper designs according to the symmetry groups they possess: Two designs would be in the same class if they had the same group of symmetries. Fedorov showed that there are exactly 17 possible symmetry groups of wallpaper designs. In the pages of this text on wallpaper designs, you will notice that there are 17 designs, one for each of Fedorov's wallpaper symmetry groups. There are representatives of many of these classes in the decorations of the ancient Egyptians and Chinese. In the ornaments decorating the walls of the Alhambra at Granada, there are representatives of all 17 classes.

Fedorov was really interested in enumerating the classes of symmetry groups of crystals. According to his definition, crystals are the three-dimensional analogues to two-dimensional wallpaper designs. Thus, a crystal is a shape in space that has translational symmetry in at least three non-coplanar directions and a smallest nonzero translational symmetry in any one direction. This is a formal definition: When you or I hold a "crystal" in our hand, we are holding only a piece of the infinite crystal defined. Fedorov accomplished his project in 1890. The 230 crystallographical space groups he enumerated form the basis of modern research on the structure of matter.

✳ *References*

Baglivo, J. A., and J. E. Graver, *Incidence and symmetry in design and architecture.* New York: Cambridge University Press, 1983.

Coxeter, H. S. M., *Introduction to geometry.* New York: Wiley, 1969.

Fejes Tóth, L., *Regular figures.* New York: Macmillan, 1964.

Guggenheimer, H. W., *Plane geometry and its groups*. San Francisco: Holden-Day, 1967.

Hilbert, David, and S. Cohn-Vossen, *Geometry and the imagination*. New York: Chelsea Publishing Co., 1952.

Holden, Alan, *Shapes, space and symmetry*. New York: Columbia University Press, 1971.

Kappraff, Jay, *Connections: The geometric bridge between art and science*. New York: McGraw-Hill, 1991.

Kline, Morris, *Mathematical thought from ancient to modern times*. New York: Oxford, 1972.

Martin, G. E., *Transformation geometry: An introduction to symmetry*. New York: Springer-Verlag, 1982.

O'Daffer, Phares, and Stanley R. Clemens, *Geometry: An investigative approach*. Menlo Park, CA: Addison-Wesley, 1997.

Schattschneider, Doris, "The plane symmetry groups: Their recognition and notation." *American Mathematical Monthly*, June–July 1978, 439–450.

Schattschneider, Doris, *Visions of symmetry: Notebooks, periodic drawings, and related work of M. C. Escher*. New York: W. H. Freeman, 1990.

Shubnikov, A. V., and V. A. Koptsik, *Symmetry in science and art*. New York: Plenum Press, 1974.

Senechal, Marjorie, and George M. Fleck, *Patterns of symmetry*. Amherst: University of Massachusetts Press, 1977.

Steen, Lynn Arthur, ed., *On the shoulders of giants*. Washington, D.C.: National Academy Press, 1990.

Weyl, Hermann, *Symmetry*. Princeton: Princeton University Press, 1952.

Yale, Paul B., *Geometry and symmetry*. San Francisco: Holden-Day, 1968.

7 *Which Shapes Are Best?*

In Chapter 4, we looked for shortest paths satisfying certain constraints. In this chapter we will consider other classes of optimization problems. We will look for two-dimensional shapes, satisfying certain constraints, having smallest perimeter or biggest area.

For example, consider the following two problems.

- You want to build a rectangular garden enclosed with a fence. With your budget you can buy at most 300 ft of fencing. You want to grow as much as you can in the garden, so you want the area of the garden to be as large as possible. What should the dimensions of the rectangle be?

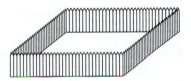

- You operate a tuna canning factory. You pack tuna in cylindrical cans of the traditional size: height 3.5 cm, radius 4.25 cm. Each can contains about 200 ml of tuna. You are wondering if this particular cylindrical shape is the best choice. Is there one that holds the same amount of tuna but uses less metal?

These are both problems of the following kind: Of all shapes satisfying a certain constraint, is there one which has the largest area? Or volume (if appropriate)? Which one? Or smallest perimeter? Usually there is a realistic context in which the answer to each of these problems might be important—such as to maximize benefits or to minimize costs.

Our initial shapes will be rectangles in the plane. Our problem-solving experience will lead us to ask: What are some analogous problems? The several responses to this question will take us in several different directions; each of these, in turn, will branch out in other directions. But the question that holds these problems together is: Which shape is best?

The words maximize and minimize, to anyone who has studied calculus, suggest (in the style of Pavlov's dog) a certain routine set of activities (find a function, calculate derivative, set equal to zero, and solve). That approach will work in some cases, but not all. In fact, an interesting feature of this chapter is the variety of solutions that will come up—some algebraic, some geometric, some just down-right clever.

Some benefits to all of this is that we get to see some familiar shapes in a new light because they are the **best** shape for certain situations. Maybe this is why the shapes are so familiar in the first place.

※ The Garden Problem

Here again is the first problem:

You want to build a rectangular garden enclosed with a fence. With your budget you can buy at most 300 ft of fencing. You want to grow as much as you can in the garden, so you want the area of the garden to be as large as possible. What should the dimensions of the rectangle be?

 STOP! Think about this before reading on. Draw pictures. Make a guess. Solve it if you can. Use any approach you feel comfortable with.

*

A Start

Surely we'll want to use up all the fencing we can buy (agreed?): 300 ft of it. So our garden will be a rectangle having a perimeter of 300 ft. Let's draw a picture of it.

Top view of garden

Something missing? Yes. Length and width don't have numbers. To get started (and get a feel for the problem) let's make up numbers for length and width that work.

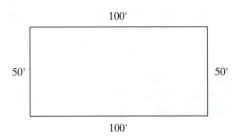

Length = 100 ft, width = 50 ft, area = 100 × 50 = 5000 ft². Good start. Are there other numbers that would work? How about length = 140, width = 10, area = 1400 ft².

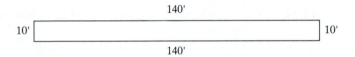

Two possibilities. Two different areas. There's a **real** problem here!

Problem 1   We've made a couple of guesses: $L = 100$, $W = 50$; $L = 140$, $W = 10$. Try continuing with this approach. Be a bit more systematic. Set up a table like the following:

Width	Length	Area
10	140	1400
20	?	?
30		
40		
50	100	5000

 Fill in the missing values. Add some more. Look for patterns. Make conjectures. (This would be a good problem for exploration with Geometer's Sketchpad.)

Solution 1.   If you have been thinking calculus, you may also have been wondering "What's going on here? Why is he fooling around drawing all these pictures? Let's assign some variables and get on with it."

The rectangle has length L and width W. We want to know what L and W should be in order that the area A be largest. What do we know about L and W relative to the rectangle? We know the rectangle has perimeter 300. We also know that the perimeter is $2L + 2W$. So we have an equation:

$$2L + 2W = 300$$

We also know that $A = LW$. So two equations:

$$2L + 2W = 300 \quad \text{and} \quad A = LW$$

We want to maximize A. In calculus we would want to take the derivative of A and work from there. To do this A has to be a function of a single variable. We can make this happen by solving for L (say) in $2L + 2W = 300$ to get

$$L = 150 - W$$

and plug this into the equation for A and get

$$A = LW = (150 - W)W = 150W - W^2$$

a function of a single variable. Differentiating this function, we get

$$A' = 150 - 2W$$

Now, $A = 150W - W^2$ has the largest value when the derivative is equal to zero. So we want to solve the equation

$$A' = 150 - 2W = 0$$

The solution is $W = 75$. When $W = 75$, $L = 75$ also, and $A = 75 \times 75 = 5625$ sq ft.

Problem 2   The reasoning in Solution 1 is incomplete. What do you need to add to finish it up?

Solution 2.   Using calculus is o.k., but it's not really necessary. In Solution 1, before using calculus, we obtained the equation $A = (150 - W)W$, which we recognize as the equation for a parabola. The parabola opens downward and has roots 0 and 150 and so has the following graph.

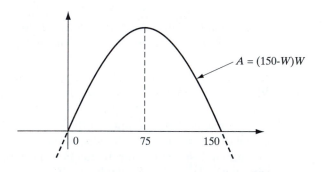

Now, the "peak" (or vertex) of the parabola always occurs exactly halfway between the two roots. In this case, the peak occurs when $W = 75$. That solves the problem.

Solution 3. Here is a third solution that looks more "geometric," more like what the Greeks might have done. Before we describe it, let's consider the mathematical form of the garden problem:

Of all rectangles having perimeter 300, which one has the greatest area?

This has a simple, natural generalization:

Of all rectangles having fixed perimeter P, which one has the greatest area?

We'll solve the latter.
    Construct a rectangle of length *L*, width *W* and perimeter *P*.

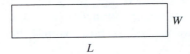

Along one side of this rectangle and with a vertex in common, construct a square also having perimeter *P*.

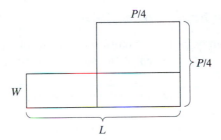

(Of course, if the rectangle is a square to begin with, then the rectangle and square coincide. Let's assume the rectangle is not a square and that $W < L$.)
    We want to show that the area of the square is bigger than the area of the nonsquare rectangle. This, in turn, will show that the area is maximized when the rectangle is a square. To show that the area of the square is bigger than the area of the rectangle really amounts to showing that the area of the portion shaded ///// is bigger than the area of the portion shaded ▓▓▓. (Why?)

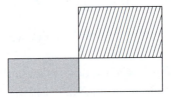

To do this, first observe that a side of a square of perimeter P must be $P/4$. Then let's label the diagram.

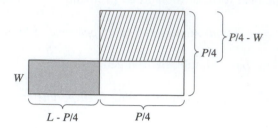

Thus, the area of the portion shaded ///// is equal to $(P/4)(P/4 - W)$. The area of the portion shaded ▬▬▬ is equal to $W(L - P/4) = W(P/2 - W - P/4) = W(P/4 - W)$. Since $W < P/4$ (why?), the latter area is less than the former. This is what we wanted to show, namely, that

> *Of all rectangles of fixed perimeter, the square and only the square has largest area.*

✳ *A Purely Algebraic Formulation*

We can write the conclusion to our problem symbolically. The area of the rectangle is LW and the area of the square is $(P/4)^2$. We have just shown that

$$LW \leq (P/4)^2 \text{ with equality occurring exactly when } L = W$$

But, since $P/4 = (L + W)/2$, this can be rewritten as

$$\sqrt{LW} \leq (L + W)/2 \text{ with equality exactly when } L = W$$

Notice that this last statement is one about numbers L and W. It doesn't rely on the interpretation that L and W are the sides of a rectangle. It does need the hidden assumptions $L > 0$ and $W > 0$. With the geometric connotations removed, our Theorem (yes!) becomes the following.

Theorem 1. If $x > 0$ and $y > 0$ are two numbers, then

$$\sqrt{xy} \leq (x + y)/2 \text{ with equality exactly when } x = y$$

Solution 4. What follows is a purely algebraic proof of the theorem and hence another solution to the original problem.

First, notice that the following statement (for positive numbers x and y)

$$\sqrt{xy} \le (x + y)/2 \text{ with equality exactly when } x = y$$

is equivalent to the statement

$$[(x + y)/2]^2 - xy \ge 0 \text{ with equality exactly when } x = y$$

To prove the latter statement, we notice that

$$[(x + y)/2]^2 - xy = (x + y)^2/4 - xy$$
$$= (x^2 + 2xy + y^2)/4 - xy$$
$$= (x^2 + 2xy + y^2 - 4xy)/4$$
$$= (x^2 - 2xy + y^2)/4$$
$$= (x - y)^2/4$$

Since the latter is the square of a number it must always be greater than or equal to zero. Furthermore, $(x - y)^2/4 = 0$ exactly when $x = y$. Thus, we have shown

$$[(x + y)/2]^2 - xy \ge 0 \text{ with equality exactly when } x = y \qquad *$$

Problem 3

For positive numbers x and y the expression $\sqrt{xy}$ is called the **geometric mean** of x and y; the expression $(x + y)/2$ is called the **arithmetic mean** of x and y. The arithmetic mean is what we get when we take the **average** of the numbers. In this problem you will show that the geometric mean can also be useful.

You want to find the weight w of an object. If you have a pan balance, then to figure out w you normally balance the object with a known weight. This depends on having the arms of the balance equal. Here is a method for finding w using a pan balance in any case, arms equal or not. Let the lengths of the arms be L and M. (You don't need to know what these numbers are!)

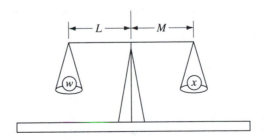

Put the object (of unknown weight w) on the left pan and balance it on the right pan with known weight x. Then put the object on the right pan and balance it on the left with known weight y. By the law of the lever, you have $wL = xM$ and $yL = wM$. Show that w is equal to the geometric mean of x and y.

Problem 4   You plan to build a rectangular pasture with one side along an existing barn. You plan to use exactly 300 ft of fencing. You want to know how to build the pasture so that its area is greatest.

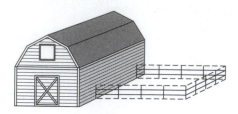

(Geometer's sketchpad might be useful here.)

* *The Storage Problem*

You are building an outdoor storage area for lumber and tools. You have decided to make it in the shape of a rectangle and you need an area of 900 ft². To keep out vandals and animals, you plan to surround the area by a fence. You want to know what the dimensions of the rectangle should be in order to minimize the cost of the fencing. (Does it make any difference?)

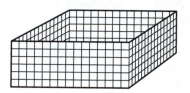

!   *STOP! Before reading on, try to solve this problem yourself. Draw a picture. Make a guess. Try getting started as we did for the Garden Problem.*

*   *

A Solution to the Storage Problem.   You have a rectangle. Its area is 900 sq ft. You buy fencing by the linear foot. So the cost of the fencing is some fixed dollar amount times the length of the fencing. The length of the fencing is the perimeter. Thus, to minimize the cost of the fencing all you have to do is minimize the perimeter. You think of the Garden Problem and its solution, but you detect a difference. In the Garden Problem the **perimeter was fixed** and you wanted to **maximize the area**. In the Storage Problem the **area is fixed** and we want to **minimize the perimeter**. Let's have a look at the inequality anyway to see if it might be useful:

$$\sqrt{xy} \le (x + y)/2 \text{ with equality exactly when } x = y$$

The numbers x and y are the length and width of the rectangle. Can any other parts of the inequality be interpreted? We recall that the right-hand side is a quarter of the perimeter. What's the left-hand side? Yes! The square root of the area. We can rewrite it in the following way.

$$\sqrt{A} \leq P/4$$

Where A is the area of the rectangle and P is its perimeter. Now A is fixed (it's 900 sq ft). We want to make P as small as we can. But P is constrained by the inequality. It must always be bigger than or equal to $4\sqrt{A}$ (or 120 ft in our case). So the smallest P can be is exactly $4\sqrt{A}$. According to the theorem, this happens exactly when $x = y$. (For our problem, this means that $P = 2x + 2y = 120$ or $x = y = 30$.)

We have just proved the following.

Of all rectangles with fixed area, the one (and the only one) with the least perimeter is the square.

✳ More Garden, Storage and Building Problems to Solve

Problem 5   As we did for the Garden Problem, solve the Storage Problem in two other ways: (1) using calculus; (2) by graphing. Comments?

Problem 6   You are a supplier of wholesale building materials. You are thinking of making a bid to be the supplier for a country cabin having the following specifications:

The cabin will be simple, consisting of a single rectangular room having 800 sq ft of living space. The cabin will sit on a cement slab. (The building site is level.) The cement slab will sit on footers 12″ wide and 16″ deep around the periphery of the slab. The cabin will be 8 ft tall. The roof will be flat and made of tin. There will be two solid pine doors, each 3 ft × 7 ft (including framing). There will be eight windows, each 2 ft × 4 ft (including framing). The walls will be made of adobe brick, each 12″ × 12″ × 3″. The bricks should be laid on top of the slab, on the part just above the footers.

You have before you a list of wholesale prices for the items that will be needed to build the house:

- pre-mixed cement: $2/sq ft of slab (the slab will be 4″ thick)
- tin roofing: $3/sq ft (the roofing comes in sheets which can be pre-cut to size)
- 3′ × 7′ solid pine door: $50 (hardware, framing, etc., included)
- 2′ × 4′ window with wooden sash and frame, no frills: $30 (hardware included)
- 12″ × 12″ × 3″ adobe brick: $1 each.

Of course, you want your bid to be lowest so that the owner will select you to be the supplier for the cabin. (In order to make your bid the lowest possible, you should try to identify features of the cabin that could vary in cost.)

Write up your bid in the form of a business proposal to the owner. In this proposal, you will want to explain your bid. Itemize costs in a way that he or she will understand. Convince him or her that these costs are necessary given the constraints of the specifications above. You will also want to convince the owner that your bid is the lowest possible. You might want to back up your arguments with charts and diagrams.

Problem 7 You have four straight, rigid pieces of fencing to surround your garden. Two of them are 10 ft long; two are 15 ft. Since you'd rather not cut them, you decide to make your garden in the shape of a parallelogram, 10 ft on one side and 15 ft on the other. Given this constraint, what should you do to make the area of your garden as big as you can? (A neat way to model this situation is to use D-Stix. Take two sticks of one length, two of another, and four connectors. Assemble a parallelogram and flex it to see when the area is biggest. Alternatively, you might want to use Geometer's Sketchpad.)

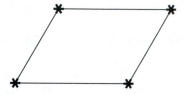

Problem 8 You are pondering your garden again. You are still limited to 300 ft for the surrounding fence. But you think: "What if I allowed the shape of the garden to be a parallelogram, would I be able to make it have greater area than if I were to limit its shape to a rectangle?"

Problem 9 Once again you are building a pasture against the side of a barn. As before, you have 300 ft of fencing. This time the pasture is to be triangular shaped, with one side of the triangle lying along the side of the barn. (You won't need fencing for that side.) Again, you want the pasture to have the largest area possible.

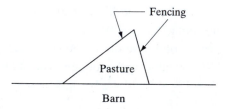

Problem 10   This time you are building a triangular pasture in the corner of an existing pasture. You will be using exactly 300 ft of fencing. You want to know how the fencing should be placed in order to maximize the new pasture's area. Use Geometer's Sketchpad to explore the possibilities. Then solve the problem.

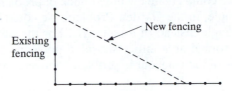

Problem 11   You and a neighbor are planning to plant gardens side by side as in the picture, with fencing all around and dividing the two plots. The two of you together can afford a total of 100 ft of fencing. What should the dimensions of the big plot be in order that the area of the two gardens be as large as possible? Geometer's Sketchpad might help you to investigate the possibilities here.

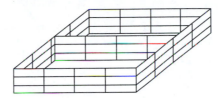

Problem 12   Three business persons—Jones, Garcia, and Wu—are constructing a joint outdoor storage area, as in the picture. They want to buy fencing to surround the entire rectangular area as well as to keep items from the three businesses separated. The total storage area will be 900 ft². The three compartments are to have equal areas. They want to know what the length and width of the total area should be in order to use the least amount of fencing. Find out.

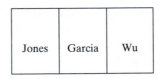

Top view of storage area

Problem 13   (This is also problem 31 of Chapter 4.)
(a) Of all triangles having fixed area A and fixed length of one side, which one has the sum of the remaining two sides smallest?
(b) Of all triangles having one side fixed and the sum of the other two sides fixed, which one has the largest area?
Geometer's Sketchpad might be useful here.

* *Related Problems and Next Steps*

We've been successful in solving the Garden and Storage Problems! Maybe we can use the techniques to solve similar problems. Let's consider some.

By replacing the rectangle with fixed perimeter by other shapes with fixed perimeter we could create a whole bunch of problems similar to the Garden Problem. For example, in the Garden Problem, we could decide that we want the garden to be a triangle instead of a rectangle. For the Storage Problem, we make the shape of the storage area into a six-sided polygon. There are many problems of this sort. Some of them appear in problems above, some below. Others we will consider later.

Problem 14   Of all triangles with a given perimeter, which one has the maximum area? (Compare this with problem 13.)

Problem 15   Of all triangles with fixed base and inscribed in a fixed circle, which one has the greatest area?

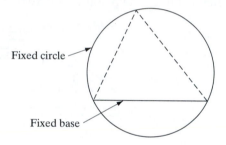

The Garden and Storage Problems suggest the two-dimensional problems just mentioned. They might also suggest three-dimensional problems. A three-dimensional analogue to a rectangle is a rectangular box. Volume is analogous to area; surface area is analogous to perimeter. So a problem analogous to the Garden Problem might read like this:

Of all rectangular boxes with fixed surface area,
which one has the largest volume?

A companion problem, a problem analogous to the Storage Problem, is the following:

Of all rectangular boxes with fixed volume,
which one has the smallest surface area?

Problem 16   What are some more three dimensional problems of this kind? (Hint: Vary the three-dimensional shape, just as we varied the two-dimensional shape in creating the two-dimensional variations of the Garden and Storage Problems.)

Problem 17   [From Pólya's (1954) *Induction and Analogy*] Of all tetrahedra inscribed in a fixed sphere, which one has the maximum volume? (Do you know a related problem?)

✳ *A Theorem*

Before trying to solve some of these problems, there is another feature of our discussion worth considering. The problems we have just accumulated are geometric variations and analogues to the Garden and Storage Problems. You recall that the inequality relating the arithmetic and geometric means is a purely algebraic one. An algebraic form of the Garden Problem leading to that inequality might have looked like this:

> *Out of all pairs of positive numbers* x *and* y *whose sum is fixed, which pair has the biggest product* xy?

Perhaps you thought of the following generalization of the problem:

> *Out of all triples of positive numbers* x, y, *and* z *whose sum is fixed, which triple has the biggest product* xyz?

Let's think about this problem. Suppose x, y, and z are positive numbers whose sum is S. The solution to the Garden Problem suggests that the product is biggest when $x = y = z = S/3$. Let's try to prove this.

Suppose that out of all collections of three positive numbers having sum S, there is one—a, b, c—whose product abc is biggest. We claim first that any pair of numbers selected from a, b, c must be equal. If we can do this, we'll be done because if $a = b$ and $b = c$, then $a = b = c$. We'll show that $a = b$. The argument for $b = c$ would be the same.

We'll prove $a = b$ by contradiction. So assume a and b are not equal. If $a + b = R$, then $ab < (R/2)^2$ from what we did with the Garden Problem or one of its variations. Thus, also

$$abc < (R/2)(R/2)c$$

But $R/2$, $R/2$, c is also a collection of three positive numbers whose sum is S. Consequently, the latter inequality contradicts our assumption that a, b, c has the biggest product. This contradiction shows that a and b must be equal. A similar argument shows that $b = c$.

Thus, we have shown that $a = b = c$. Since also $a + b + c = S$, we must have $a = b = c = S/3$.

The biggest product must be $(S/3)^3$.

Here is what we have proved. Suppose that x, y, z are positive numbers whose sum is S. Then

$$xyz \leq (S/3)^3 \text{ with equality exactly when } x = y = z$$

If we take cube roots of both sides of the inequality and replace S by $x + y + z$ then we get the following.

Theorem 2.   Suppose that x, y, z are positive numbers. Then

$$\sqrt[3]{xyz} \le (x + y + z)/3 \text{ with equality exactly when } x = y = z$$

The right-hand side of the inequality is the arithmetic mean of x, y, and z, and the left-hand side is the geometric mean. This theorem is a neat generalization of the theorem we proved earlier in conjunction with the solution to the Garden Problem.

[**Please note:** Our proof of this theorem—the Theorem of the Arithmetic and Geometric means for three positive numbers—is missing a piece. Among all the products xyz of three positive numbers with fixed sum, we **assumed** there was one (or more) that was biggest. A complete proof of the theorem must spend some time proving the **existence** of three positive numbers x, y, z such that xyz is biggest. We won't deal with this here.]

Problem 18

In thinking about some problems we have solved, the following question may have occurred to you:

Out of all collections of n *positive numbers* x_1, x_2, . . . , *and* x_n *whose sum is fixed, which collection has the biggest product* $x_1 x_2$ . . . x_n?

The theorem just stated answers this in case n = 3; an earlier theorem answers this for $n = 2$. What would be a generalization of the two theorems? (The theorem should provide an answer to the question just posed.) Prove the theorem assuming the existence part, as referred to in the note above.

✳ *Using the Theorem to Solve Problems*

The Theorem of the Arithmetic and Geometric Means for two positive numbers says the following.

Suppose that x *and* y *are positive numbers. Then* $\sqrt{xy} \le (x + y)/2$ *with equality exactly when* x = y.

We proved this earlier and used it as a solution to the Garden and Storage problems where xy is the area of a rectangle and $x + y$ is half the perimeter of the rectangle. Earlier we talked about three-dimensional analogues to these problems. Maybe the Theorem of the Arithmetic and Geometric Means for three positive numbers could help us solve them. Recall what it says.

Suppose that x, y, *and* z *are positive numbers. Then* $\sqrt[3]{xyz} \leq (x + y + z)/3$
with equality exactly when x = y = z.

If we are going to use this to solve problems, it's natural enough to think of *xyz* on the left-hand side of the inequality as the volume of a rectangular box having length, width, and height equal to *x*, *y*, and *z*. On the right-hand side of the inequality the expression $x + y + z$ appears. What does that measure? The following is a problem involving a box, its volume, and this sum.

✳ The Post Office Problem

You are interested in shipping plastic noodles by mail. You want to use rectangular boxes. The Post Office limits the size of a rectangular box that it will ship. It requires that the sum of length, width, and height of the box must not exceed 108″. You want to ship as many plastic noodles in a box that you can. So you want to know what the shape of the box should be that not only satisfies this requirement but also has the greatest volume.

! *STOP! Try this yourself before reading on.*

* *

A Solution to the Post Office Problem. You assume that your box has $x + y + z = 108''$. Thus, $(x + y + z)/3 = 108''/3 = 36''$. Thus, also

$$\sqrt[3]{xyz} \leq 36$$

by the Theorem of the Arithmetic and Geometric Means (for three positive numbers). In other words,

$$xyz \leq 36^3$$

Thus, the biggest *xyz* can be is 36^3. By the same theorem $xyz = 36^3$ exactly when $x = y = z = 36$. To have maximum volume, your box must be a cube with 36″ to a side.

Problem 19   I lied. Here is what the Post Office really said: "The longest linear measurement plus the girth (distance around the middle) must be less than 108 inches." Is this really different than what was stated in the Post Office Problem? Would the answer to this problem be any different than the solution to the Post Office Problem? (You might want to take a peek right now at problem 23.)

The Post Office Problem got us warmed up in the use of the Theorem of the Arithmetic and Geometric Means, but it's not quite what we had in mind as an analogue to the Garden or Storage Problems. One of the three-dimensional problems we set out earlier was the following.

Of all rectangular boxes with fixed volume,
which one has the smallest surface area?

Let's give it a realistic context.

✳ *The Cereal Box Problem*

You are the manufacturer of Sawzy-Dusties[R], a "natural" cereal. You distribute this cereal in rectangular boxes each having volume 3375 cm³. You want to know what the shape of the box should be in order to minimize the amount of cardboard used in making it.

! ***STOP! Try this before reading on. Draw pictures. Make guesses.***

(A good way to get a real feel for this problem is to make some of the possible cereal boxes. Since 3375 = 3 · 3 · 3 · 5 · 5 · 5 = $3^3 5^3$, some possible boxes are 9 × 15 × 25 and 5 × 27 × 25 and 3 × 3 × 375 and 15 × 15 × 15. If you make these boxes—and others like them having volume 3375 cm³, you can do a few things: (1) line them all up, look at them, and see if you can guess which one uses the least amount of cardboard (maybe you'll think they all use the *same* amount; maybe making the boxes will suggest a ranking of the amounts of cardboard . . .); (2) calculate the surface area for each one and compare; (3) see if there is any correlation between what you get from (1) and (2).)

* * *

A Solution to the Cereal Box Problem. Two featured ingredients in the problem are the box's volume and the amount of cardboard. We assume the latter is measured by the box's surface area.

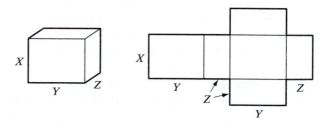

If the dimensions of the box are the positive numbers x, y, and z, then xyz is the box's volume and $2xy + 2xz + 2yz$ is the surface area. The quantity xyz appears on the left-hand side of the inequality of the theorem. But where in the inequality is the surface area? The right-hand side of the inequality is perfectly suited for the Post Office problem, but not this one. That's a real pickle. What to do?

Let's look at things from another perspective. The surface area $2xy + 2xz + 2yz$ is a **sum** of positive numbers. What would happen if we used the three numbers $2xy$, $2xz$, and $2yz$ in the theorem? Their product is $(2xy)(2xz)(2yz) = 8(xyz)^2$. According to the theorem, we have

$$\sqrt[3]{8(xyz)^2} \le (2xy + 2xz + 2yz)/3 \text{ with equality exactly when } 2xy = 2xz = 2yz$$

If V is the volume of the box, then the left-hand side is $2V^{2/3}$, a fixed number ($V = 3375$). So the surface area, $2xy + 2xz + 2yz$, must always be greater than or equal to $6V^{2/3}$. So the **smallest** the surface area can be is $6V^{2/3}$. This happens exactly when

$$2xy = 2xz = 2yz$$

i.e., when $x = y = z$, i.e., when the box is a cube.

The moral to the story: The Theorem of the Arithmetic and Geometric Means is a powerful tool. But it doesn't allow you to turn off your mind. You've got to use your noodle. You will sometimes need to be clever in solving the following problems.

Problem 20   Most cereal boxes are not cubes. Why is this?

Problem 21   In the solution to the cereal box problem, we showed that for any rectangular box with sides x, y, and z

$$2V^{2/3} \le S/3 \text{ with equality exactly when } x = y = z$$

where V is the volume of the box and S is its surface area. We used this essentially to show that of all boxes with a given volume, the cube has minimum surface area. Use it to solve the companion problem:

Of all rectangular boxes with fixed surface area, which one has the greatest volume?

✳ *Yet More Packaging and Building Problems to Solve*

Problem 22   A few years ago the International Air Transport Association (IATA) changed the individual baggage limitation (on international—not domestic—flights). Previously, each person was limited to 20 kg. Now the limitation is the following.

- First class. Two suitcases allowed. Each must satisfy

$$\text{length} + \text{width} + \text{height} \leq 62''$$

- Second class. Two suitcases allowed. Each must satisfy the restrictions for a suitcase in first class plus the sum of the two lengths, two widths, and two heights must not exceed 106″ (!).

For each class, design rectangular-box-shaped luggage pieces so that the requirements are satisfied and the total volume is biggest.

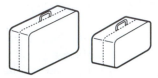

Problem 23

You have decided to start shipping your plastic noodles in cylindrical boxes. As before, the Post Office limits the size of a such a box for shipping. It requires that the sum of the height and the distance around (the curved part) of the box must be less than or equal to 108″. Again you want to ship as many plastic noodles in a box that you can. So you want to know what shape box satisfies this requirement and has the greatest volume. Find out.

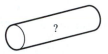

Problem 24

(This is the Tuna Can Problem stated at the beginning of the chapter.) You operate a tuna canning factory. You pack tuna in cylindrical cans of the traditional size: height 3.5 cm, radius 4.25 cm. Each can contains about 200 ml of tuna. You are wondering if this particular cylindrical shape is the best choice. Is there one that holds the same amount of tuna but uses less metal?

Problem 25

You are designing a litter barrel in the shape of a rectangular box without top. You have selected the material for the can and have determined what the weight of the can should be. This will limit the surface area to 2 m². You want to know what the dimensions of the box should be for the can to hold the maximum volume of litter.

Problem 26   You are also considering a cylindrical shape as an alternative for the litter barrel mentioned in problem 25. Again the surface area will be limited to 2 m². Again you want to know what the height and radius should be for the barrel to hold the greatest volume of litter.

Problem 27   You are designing a new two-person tent called the "Tepee." It will be made of canvass, have a conical shape, and be without a floor. You have scientifically determined that the tent must have a volume of 3 m³. You want to know what the height and radius of the cone should be in order to minimize the amount of material used in the tent.

Problem 28   Recall the Cereal Box Problem. We assumed that the amount of cardboard was measured by the surface area. This assumes that there is one layer of cardboard throughout the whole surface. A more realistic assumption is that the top, bottom, and one side form a double thickness of cardboard and that the remaining sides are single thickness. The volume of the box is as before. With these assumptions, what should be the dimensions of the box for the amount of cardboard to be minimal?

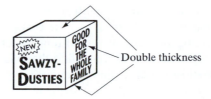

Problem 29   An air-conditioned storage building is to be constructed in a warm climate. Its shape is a rectangular box with a volume of 75,000 ft³. No heat comes into the building through the floor. However, heat coming into the building through the walls is five times greater per square foot than that through the roof. What should the dimensions of the building be in order to minimize the amount of heat entering the building?

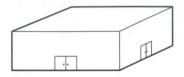

Problem 30   You are making the trough below out of tin. You can only afford 20 ft² of tin. How big can the volume of the trough be? And what should the measurements of the trough be when its volume is biggest?

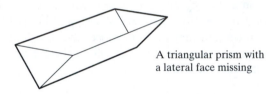

A triangular prism with a lateral face missing

Problem 31   Take a pyramid with square base. Assume that a perpendicular from apex to base passes through the center of the square. (Such a pyramid is called a **right** pyramid.) Put two of these together to make a **double pyramid**. Of all such double pyramids with fixed volume, which one has the smallest surface area?

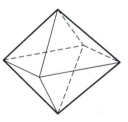

Problem 32   In problem 24, 200 ml of tuna is to be packed into a cylindrical can. A reasonable assumption about the construction of the tuna can is that the cost per cm² for circular top and bottom and the cost per cm² for the lateral sides are different. Assuming the cost of top/bottom material is $.001/cm² and the cost of lateral side material is $.0005/cm², find the height and radius of the can that minimizes the cost of materials in making it.

Problem 33   You own a grocery store and are building a large rectangular box from which to sell bulk foods. The box will have several compartments, as shown in the picture below. The box and the dividers between the compartments will be made out of plexiglass. (For the moment you may want to assume that the thickness of the plexiglass is negligible.) There will be 15 compartments; the volume of each compartment is to be 20 liters. What should the dimensions of the box and the compartments be in order to minimize the cost of materials used to make box and dividers? (The lid—the top of the box—will be hinged so that compartments can be filled. A gadget will be attached to the front of each compartment through which bulk food from within can flow into a customer's bag.)

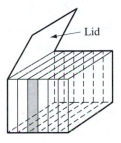

Problem 34   The following problem is found in most calculus books. Can you solve it using the Theorem of the Arithmetic and Geometric Means?

You have a rectangular piece of cardboard whose length is S inches and width is T inches. You plan to make an open box out of the cardboard by cutting out square corners, folding up the sides and taping. You know that you have a choice as to the size of the corners you cut out, but you want to know which size will give you the biggest volume.

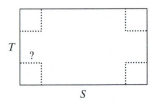

Problem 35   Of all tetrahedra with a fixed volume, which one has the smallest surface area?

✻ *The Isoperimetric Problem*

The Garden Problem suggested several related two-dimensional problems, some of which we considered, while others were just mentioned. Here are a couple of examples.

> *Of all triangles with fixed perimeter, which one has the greatest area?*

Here is its companion:

> *Of all triangles with fixed area, which one has the smallest perimeter?*

Then there is a problem close to the original Garden Problem:

> *Of all quadrilaterals with fixed perimeter, which one has the greatest area?*

And here is its companion, a problem close to the Storage Problem:

> *Of all quadrilaterals with fixed area, which one has the least perimeter?*

These two pairs of problems generalize to two larger families of problems one of which is the following.

> *Of all* n-*gons with fixed perimeter (and fixed* n*), which one has the greatest area?*

Here are their companions:

> *Of all* n-*gons with fixed area (and fixed* n*), which one has the least perimeter?*

We'll have a look at these problems later. There is one additional problem that is the mother of all problems of this kind:

> *Of all two-dimensional shapes with fixed perimeter, which one has the greatest area?*

This problem also has a companion:

> *Of all two-dimensional shapes with fixed area, which one has the least perimeter?*

The first of these two problems is called the **Isoperimetric Problem**, isoperimetric meaning "same perimeter." (Of all shapes having the same fixed perimeter, . . .) You probably suspect what the answer to both problems are. Descartes, in considering the second problem, calculated the perimeters of several shapes all having the same area (1 sq in). He assembled the following data:

Figure	Perimeter (in.)
Circle	3.55
Square	4.00
Quadrant of circle	4.03
Rectangle 3:2 (*)	4.08
Semicircle	4.10
Sextant of circle	4.21
Rectangle 2:1	4.24
Equilateral triangle	4.56
Rectangle 3:1	4.64
Isosceles right triangle	4.84

*Rectangle 3:2 means a rectangle whose sides are in the ratio 3 to 2.

Problem 36

(1) Find the perimeter of a regular hexagon having area 1. (2) Find the perimeter of a regular octagon having area 1.

If Descartes had known what we know, he wouldn't have had to calculate the perimeter for all those rectangles. In any case, the circle is the best shape among those that Descartes considered. This convinced Descartes that the circle is the best of **all** shapes. Maybe it convinces you, too. But let's produce a proof that will really clinch it for the circle. What we want to prove is the following theorem.

Same-Area Theorem. Of all shapes having the same area, the circle and only the circle has the smallest perimeter.

Of course, we also want to prove the following companion theorem.

Isoperimetric Theorem. Of all shapes having the same perimeter, the circle and only the circle has the largest area.

We will actually prove the latter. Before doing this, let's show that the Same-Area theorem is a consequence of the Isoperimetric theorem.

Proof of Same-Area Theorem Assuming the Isoperimetric Theorem. Suppose a certain shape S has area A and perimeter $\mathbf{P}$. Consider two circles, one that has area A (perimeter P and radius r) and one that has perimeter $\mathbf{P}$ (area A and radius $\mathbf{r}$). Then the areas of the two circles are also πr^2 and $\pi \mathbf{r}^2$, respectively; and their perimeters are $2\pi r$ and $2\pi \mathbf{r}$, respectively. Thus, also $P^2 = 4\pi A$ and $\mathbf{P}^2 = 4\pi A$.

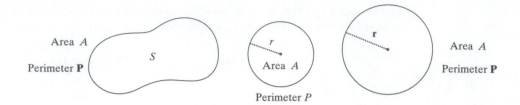

Since shape *S* has the same perimeter as the second circle, the isoperimetric theorem says that $\mathbf{P}^2 \geq 4\pi A$ with equality if and only if *S* is a circle. Consequently, $\mathbf{P}^2 \geq P^2$ with equality if and only if *S* is a circle. In other words, of all shapes with area *A* the circle (and only the circle) has least perimeter. ✻

We are almost ready to prove the big theorem. It will be helpful in the proof to know how to tell whether or not a given curve is a circle. Hence, the following problem and its solution.

Characterization of Circle Problem. In the proof of the theorem we will need some way of recognizing a circle. You recall (Chapter 2) that any angle inscribed in a semicircle is a right angle.

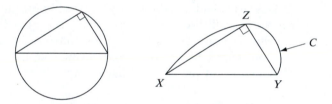

Is the converse true? In other words, suppose that a curve *C* joining two points *X* and *Y* has the property that for any point *Z* on *C*, $\angle XZY$ is a right angle. Can you then claim that *C* is a semi-circle with diameter *XY*? Let's see.

Solution to the Circle Problem. Suppose that *C* is such a curve. Rotate the curve 180° about the midpoint of *XY* to get a copy *C'* of *C*. The two copies will look something like the following.

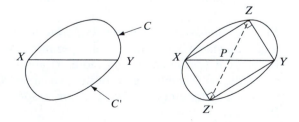

Next take a point Z on C. Call Z' the corresponding point on C'. Then $XZYZ'$ is a rectangle. One diagonal of this rectangle is XY; the other is ZZ'. Moreover the two diagonals of a rectangle meet in their midpoints P. Thus, $PZ = PX = PY = PZ'$. The argument is the same for any point on C. Thus, C coincides with a semicircle of diameter XY.

So, if we come across a curve C described in the following manner:

curve C joins two points X and Y and has the property that for any point Z on C the angle XZY is a right angle

then we'll know C is a semicircle.

It will also be helpful in the proof to have the solution to the following.

The Two-Sided Triangle Problem. Of all triangles with two given side-lengths, which one has the greatest area?

❗ **Stop! Try this before reading on.**

✳   ✳   ✳

Solution to the Two-Sided Triangle Problem. Suppose that A and B are the two side-lengths and that α is the angle between the two sides. Then the area of the triangle is $(1/2)AB \sin \alpha$.

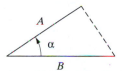

This quantity is maximum exactly when $\sin \alpha = 1$ or when $\alpha = \pi/2$.

Now we are really ready for the proof of the Isoperimetric theorem.

Proof of the Isoperimetric Theorem. We'll prove the theorem in three steps. Start with a two-dimensional shape S having perimeter P. We assume that the area of S is greatest among all shapes having perimeter P. We want to show that S must be a circle.

Step 1. We claim that S must be convex. For, if not, we can replace S by a shape whose perimeter is the same but whose area is greater. The diagrams below illustrate how this might come about.

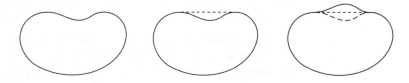

Since S already has the greatest area possible among all shapes with perimeter P, S must have been convex in the first place.

Step 2. We claim that any line through S dividing its perimeter in half must also divide its area in half.

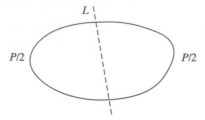

Suppose this is not true. That is, suppose that a line L divides the perimeter of S in half, but that the area to the left of L is greater than the area to its right.

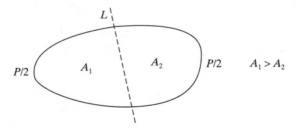

Create a second shape made up of the left-hand side of S and the mirror image of this left-hand side in the line L.

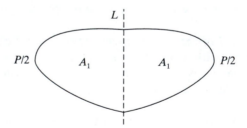

This new shape has the same perimeter as S but greater area. Since S already had greatest area, it must be that the line dividing the perimeter of S in half also divided its area in half.

Step 3. Let's take advantage of the line L that halves both the perimeter and the area of S. Ultimately we want to show that S, a shape having the greatest area among all shapes having perimeter P, is a circle. Suppose that this is not true.

Then at least one the halves of S is not a semicircle. Suppose in fact that the left-hand side is not a semicircle. Create a new shape S' made up of the left-hand side of S and its reflection in line L.

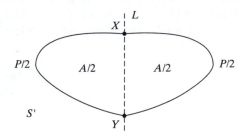

Since L halves the area and the perimeter of S, the two shapes S and S' have the same area and the same perimeter. Denote by X and Y the points where L intersects the perimeter of S'. Since the left-hand side of S (and S') is not a semicircle, there must be a point Z on the perimeter of S' such that $\alpha = \angle XZY$ is not a right angle. (See solution to Characterization of Circle Problem.) Let Z' denote the reflected image of Z in line L.

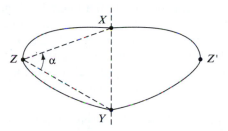

To complete the argument, we create a "contraption" out of these ingredients. The description of this contraption is unusual, so hold onto your hats. Imagine S' is a piece of cardboard. Cut out the quadrilateral $XZYZ'$. What is left are the four crescent-moon forms shaded in the diagram below. Install hinges at Z, Y, and Z' and a sleeve at X to allow X to slide along the line L. The point Y stays fixed on line L.

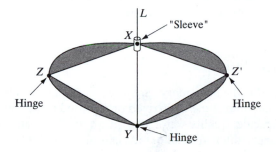

As we slide X along L we form different shapes; one of them is S. Each one has the same perimeter as S. The area of each is equal to the sum of the areas of the four crescent-moon forms plus the area of the quadrilateral they contain. As we slide X along L only the area of the quadrilateral changes. The area of the quadrilateral is equal to twice the area of triangle XYZ. The latter area is biggest when the angle at Z is a right angle. (See the solution to the two-sided triangle above.)

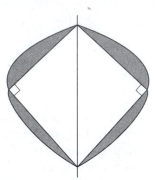

Recall that the angle at Z for shape S' is not a right angle. So the shape we create by our contraption when the angle is a right angle has greater area than S'. But this can't be! All the shapes have the same perimeter that S' has, and S' has the greatest area. Our assumption that a point Z on S' exists with angle not a right angle must be wrong. Thus, at all points of S', a right angle is created and consequently S' is a circle. So S, the original shape, must also be a circle. This completes the proof.

✳

Problem 37   If a circle of area A and perimeter P has radius r, then $A = \pi r^2$ and $P = 2\pi r$. Thus, also $P^2 = 4\pi^2 r^2 = 4\pi A$. Show that if a shape has area A and perimeter P. Then $4\pi A \le P^2$ with equality exactly when the shape is a circle.

Problem 38   The **isoperimetric quotient** of a shape with area A and perimeter P is A/P^2.
(a) Show that the isoperimetric quotient of a shape is always less than or equal to $1/(4\pi)$ with equality exactly when the shape is a circle. (See the previous problem.)
(b) The isoperimetric quotient of **any** circle is $1/(4\pi)$. What is the isoperimetric quotient of a square? (Is it the same for all squares?) What is the isoperimetric quotient of a regular hexagon? (Is it the same for all regular hexagons?)

✳ *Dido's Problem*

Dido, the legendary queen of Carthage in Virgil's *Aeneid* (and also in Purcell's *Dido and Aeneas*), arrived on the Mediterranean coast of Africa desiring to acquire upscale, quality waterfront property. The natives sold her a piece of seashore "not larger than what an oxhide can surround." Clever Dido cut the oxhide into narrow strips out of which she made a very long string. Her problem then was to figure out, with the string as perimeter, what shape would give her the greatest area. Well, not quite. The string didn't have to surround the whole piece of land, only the part of the perimeter that didn't lie along the seashore. Help Dido solve her problem.

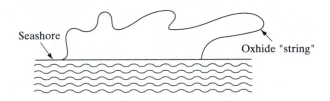

STOP! Try this before reading on.

* * *

A Solution to Dido's Problem. The picture above suggests that the seashore is a straight line of unbounded length. Let's assume that. (Problem 39 below deals with other possible seashores.) To solve the problem, reflect the curve *XYZ* across the line *XZ* to obtain a closed curve *XYZY'* of length twice that of curve *XYZ*. (Of course the length of curve *XYZ* is equal to that of Dido's oxhide strip.) The curve *XYZY'* surrounds an area twice that surrounded by curve *XYZ* and the line segment *XZ*.

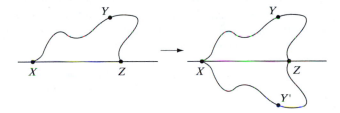

Now the problem of maximizing the area bounded by the curve *XYZ* and the straight line segment *XZ* is the same as maximizing the area bounded by the curve *XYZY'*—as long as we maintain *XZ* as a line of mirror symmetry of the curve. One such possible curve is a circle whose diameter lies on line *XZ* extended. The Isoperimetric theorem says that this closed curve of fixed length surrounds the greatest area when it is a circle with no restrictions on the diameter. This implies that the best shape for Dido's oxhide is a semicircle whose diameter lies on the line *XZ* extended. The circumference of the circle equals twice the length of the oxhide.

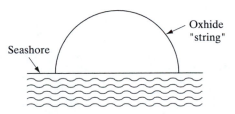

Problem 39   In the solution to Dido's Problem we assumed that the shore line was straight. What if the shore line were like one of those shown below?

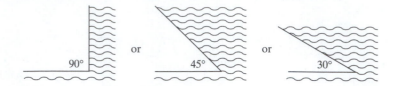

Problem 40   Generalize the results of Dido's Problem and problem 39.

Problem 41   Aeneas has 800 ft of fencing with which he can construct a rectangular pasture for his horses. The pasture will be built along the river. No fencing is needed for the river side of the pasture. Of course Aeneas wants to know what the length and width should be in order that the pasture have the greatest area possible.

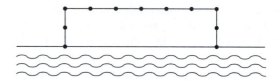

Problem 42   Aeneas is reconsidering the construction of his horse pasture. Now he thinks the pasture should be triangular. Again he wants to know which triangle will give him a pasture of greatest area. And will it be better than the rectangular solution to the previous problem?

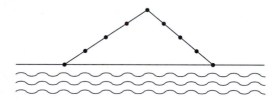

Problem 43   Aeneas now wants to build a pasture for his sheep, also along the river. He decides that the area of the pasture must be 1000 ft². Having used up all his fencing in building the horse pasture, he wants to minimize the amount of new fencing he will have to procure for the sheep pasture. What should the shape of the sheep pasture be in order to minimize the amount of fencing he'll need?

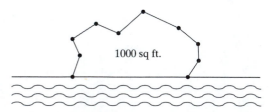

✳ *Dido's Second Problem*

Dido, having moved a bit along the coast, is still in the business of procuring waterfront property. The natives offer her "not larger than what an oxhide can surround," but this time the straight stretch of coast her land must border is a fixed length less than that of her oxhide string. She wants to know what the shape of the property must be in order that its area be maximized.

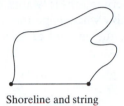

Shoreline and string

Stop! Try this before reading on.

*   *   *   *

A Solution to Dido's Second Problem. One possible configuration that string and stretch of coast could surround is a segment of a circle in which the stretch of coast is a chord and the string is an arc connecting the chord's endpoints.

[Problem 44   Given the lengths of the oxhide and stretch of coast in Dido's Second Problem, how would you construct the circle just described?]

Complete this circular segment of area *C* by adding its complementary segment of area *B*. The latter is shaded in the diagram below right.

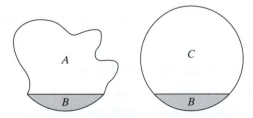

Now consider any area *A* surrounded by string and stretch-of-coast. Attach the same complementary segment of area *B* as before. The latter is shaded in the diagram on the left above. The diagram on the left above has perimeter equal to the circle on the right. Thus, by the Isoperimetric theorem, $A + B \leq C + B$. Thus, also $A \leq C$. Consequently, oxhide and stretch-of-coast surround the greatest area when the two form a segment of a circle.

Problem 45

Aeneas wants to construct another pasture along another river. This time only 50 ft of the riverside of this particular river is suitable for forming part of the pasture's perimeter. He wants the pasture to have an area of 5000 sq ft. He wants to know what shape the remainder of the perimeter should have in order to minimize the amount of fencing he will have to buy and install.

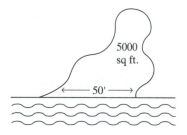

Problem 46

(Adapted from Pólya's *Induction and Analogy in Mathematics*.) You have two telephone poles (one 20 ft long, one 30 ft) and two lengths of flex-fencing (one 40 ft, one 50 ft). You want these to surround a garden of largest area. You are considering two possibilities:

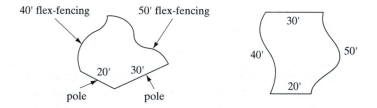

In each of these cases, how should you arrange things so that the area is greatest? Of these two arrangements, which is greatest?

Problem 47

Generalize Dido's second problem and the previous problem.

✳ *Picking the Raisins Off the Cake*

In the early part of this chapter we solved the following problems. Of all rectangles with fixed perimeter, which one has the greatest area? Of all rectangles with fixed area, which has the smallest perimeter? This suggested a whole bunch of similar problems: Of all triangles with fixed perimeter, which has largest area? Of all quadrilaterals with fixed area, which has smallest perimeter? A remarkable thing about the Isoperimetric theorem is that it can be used to solve a lot of problems like these. The way that the Isoperimetric theorem is used is to find, among all polygons having a given characteristic, one that can be inscribed in a circle. The following problem together with its solution is an example.

✳ *Garden Problem 2*

You have a bunch of logs of different lengths that you want to use to surround a garden completely. How should you arrange them so that the surrounded garden has the greatest possible area?

!

STOP! Think about this before reading on.

*

Getting Started with Garden Problem 2. Let's play around a bit. With one or two logs you can't surround much of a garden completely. With three logs, there is only one way to surround a garden.

So far there is no problem! With four logs, however, there are a lot of different ways of arranging them. Here are some:

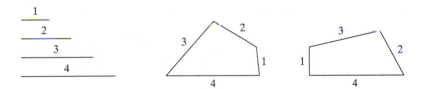

Which arrangement surrounds the greatest area? Certainly, one thing is clear: the surrounded region must be convex.

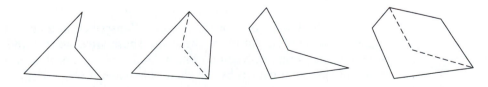

The following problems should help you think about Garden Problem 2 and gather some evidence related to it. To simplify the problem a bit, imagine that the logs are the sides of a polygon.

Problem 48

Does the order in which the different lengths are arranged around the perimeter make a difference in Garden Problem 2? Try the following concrete examples. (a) Look at all polygons with side lengths 1, 2, 1, 2 **arranged in that order**. (Draw a few.) Of all such polygons, you **know** which one P has the greatest area. (What is it and what is its area?) Next look at all polygons with side lengths 1, 1, 2, 2 **arranged in that order**. (Draw a bunch.) You also **know** which one P' of these has the greatest area. (What is it and what is its area?) How do the two areas compare? (b) Consider all polygons of side lengths 1, 2, 3, and 4 **arranged in that order**. Suppose we have one (call it P) surrounding the greatest area. Next suppose that of all polygons of side lengths 1, 3, 2, and 4 **arranged in that order** we have one (call it P') surrounding the greatest area. Which area is greater, that of P or that of P'? Or are they the same?

Problem 49

Use Geometer's Sketchpad (or some other dynamic, geometric software) to set up a polygon with sides 1, 2, 3, 4. Set it up so that the software will calculate the area of the polygon. Wiggle the vertices to obtain polygons of different areas. See if you can see which polygon gives you the greatest area.

A Solution to Garden Problem 2.   The problem we are solving is the following:

*Of all polygons with side lengths a, b, c, . . . **arranged in that order** which one has the greatest area?*

I claim that a polygon with sides $a, b, c, \ldots$ (arranged in that order) inscribed in a circle has greater area than any other polygon with sides $a, b, c, \ldots$ arranged in that order. (Given lengths $a, b, c, \ldots$ can you find a circle that inscribes a polygon with these sides? See problem 51.)

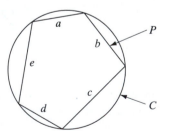

To see why this is so, take the polygon P inscribed in a circle C. Consider the segments of the circle cut off by the sides. These are shaded in the diagram below. Next, take any other polygon P' with sides $a, b, c, \ldots$ in that order. To each side attach the segment cut off by the same side when it was inscribed in the circle.

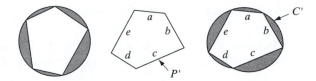

The segments of the circle, once attached to the sides of the inscribed polygon P, but still connected at what were once the vertices of P, form a shape that looks like a deflated version of its former self. The arcs of the segments now form a closed curve C' surrounding an area equal to the area of the segments plus the area of polygon P'. Since C and C' have the same length, the Isoperimetric theorem says that the area surrounded by C must be greater than the area surrounded by C'.

$$\text{area of circle } C > \text{area surrounded by } C'$$

But,

$$\text{area of circle } C = \text{area of segments} + \text{area of } P$$

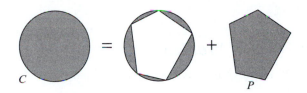

and

$$\text{area surrounded by } C' = \text{area of segments} + \text{area of } P'$$

Combining these three relationships, we get

$$\text{area of } P > \text{area of } P'$$

This is what we wanted to show. The second garden problem is solved: The arrangement of the logs which gives the greatest area is when the logs form the sides of a polygon inscribed in a circle.

We have also proved the following.

Theorem 3.   Given two polygons having the same sides (in order) with one inscribed in a circle. Then the one inscribed in a circle has the greater area.

Problem 50   Show that of all parallelograms with the same sides, only the rectangle can be inscribed in a circle. Use this and the theorem above to provide an alternative solution to problem 7.

Problem 51   We have been assuming that, given a set of lengths, there is a polygon with sides having those lengths and inscribed in a circle. Give an argument showing why this is so.

Problem 52   (Compare with problem 48.) Consider a bunch of lengths $a, b, c, \ldots$. The sides of these lengths can be ordered in many ways to make the sides of a polygon. For each ordering there is a polygon with these sides inscribed in a circle. Compare the areas of these inscribed polygons. Also compare the circles these polygons are inscribed in.

✳ *The Big Polygon Problem*

Of all polygons having n sides and perimeter P, which has the greatest area?

!

> **STOP! Think about this before reading on.**

* * * * *

Solution to the Big Polygon Problem.   Suppose we have an n-gon that has perimeter P. Suppose also that this n-gon has the greatest area among all n-gons having perimeter P. We know from what we have done earlier that this polygon can be inscribed in a circle. We also claim that all its sides are equal.

We will prove this by showing that any pair of adjacent sides must be equal. Let's show this first for the case when the n-gon is a triangle.

Take a triangle ABC having perimeter P. Suppose its area is greatest among all triangles with perimeter P. We claim that any pair of sides must be equal. To show this take sides AB and BC.

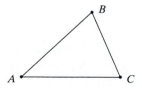

Consider all triangles having base AC and perimeter P. The third vertex of each of these triangles will lie on an ellipse with foci A and C and length sum equal to $AB + BC$.

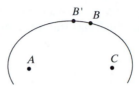

Of all these triangles (having base AC and perimeter P) the one $AB'C$ with the greatest area is the one with $AB' = B'C$. Since ABC was the one with greatest area, it must be the case that $AB = BC$ also.

Now take an n-gon S that has perimeter P and whose area is greatest among all n-gons with perimeter P. Again, we know that this n-gon S can be inscribed in a circle. We want to show that any pair of adjacent sides must be equal. To do this, look at a pair: AB and BC.

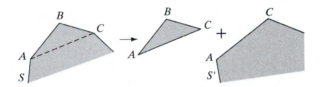

Consider the shape S' you get by slicing off the triangle ABC from S. The area of S is equal to the area of ABC plus the area of S'. If sides AB and BC are not equal, then using the argument for triangles above the triangle $AB'C$ with $AB' = B'C$ and $AB' + B'C = AB + BC$ has greater area than ABC. Gluing $AB'C$ onto S' along AC would give us a polygon with perimeter P but area greater than S. This is impossible. Thus, $AB = BC$ to start with.

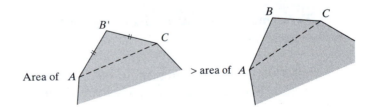

Area of A > area of A

We have just shown that any adjacent pair of edges of S must be equal. Thus, all the sides of S must be equal. Since S can also be inscribed in a circle, this means that S is a regular n-gon. We have proved the following.

Theorem 4. Of all *n*-gons with perimeter *P*, the one with greatest area is the regular *n*-gon of perimeter *P*.

Problem 53

Solve the companion problem to that solved by the theorem: Of all *n*-gons with fixed area, which has the smallest perimeter?

✳ *Which n?*

Take two regular polygons having the same perimeter, one with *n* sides and the other with *n* + 1 sides. Which has the greater area?

To answer this, we first note that the regular *n* + 1-gon of perimeter *P* has area greater than any irregular *n* + 1 gon with perimeter *P*. Next, turn the regular *n*-gon of perimeter *P* into an irregular *n* + 1-gon in the following way. We don't actually alter the *n*-gon, we just look at it differently. Think of it as an *n* + 1-gon *n* − 1 of whose sides are all equal to *L* and two adjacent sides equal to *L*/2. The angle between the two smaller sides is *π*.

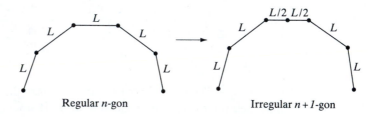

Regular *n*-gon   Irregular *n* + *1*-gon

Thus, the regular *n*-gon of perimeter *P*—an irregular *n* + 1-gon of perimeter *P*—has area less than that of the regular *n* + 1-gon of perimeter *P*. We have proved the following.

Theorem 5. A regular *n* + 1-gon of perimeter *P* has greater area than a regular *n*-gon of perimeter *P*.

Problem 54

Solve the companion problem to the theorem: Of a regular *n* + 1-gon and a regular *n*-gon both of area *A*, which has the least perimeter?

Problem 55

Aeneas is still undecided about the shape of the horse pasture along the river. He still has 800 ft of fencing. He still wants to have the area of the pasture be as great as it can and be subject to certain constraints. This time he says to himself, "Let's just make the pasture a quadrilateral with one side of it along the river." Which quadrilateral has greatest area? (Is it any better than the solutions to the earlier two problems 41 and 42?)

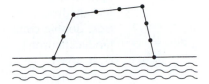

✳ *Three-Dimensional Analogue to the Isoperimetric Problem*

Having thought about the Isoperimetric theorem and some of its consequences, we naturally ask

Of all solids with fixed volume, which one has the smallest surface area?

This is called the three-dimensional Isoperimetric problem. By analogy with the two-dimensional problem, we suspect that the answer is a sphere. This conjecture is supported by the phenomenon of a spherical soap bubble, in which a certain amount of air is enclosed by a surface of minimum area. This is the physical evidence for the theorem that, **out of all solids of a given volume, the sphere (and only the sphere) has the smallest surface area**. Unlike the proof of the two-dimensional theorem, a mathematical proof of this three-dimensional theorem is difficult. The first rigorous proof was given by Hermann Amandus Schwarz in 1884. See also the Notes at the end of the chapter.

Problem 56

(a) Assuming that the three-dimensional Isoperimetric theorem is true, show that for a solid of volume V and surface area S the following inequality holds:

$$V^2 \leq S^3/(36\pi)$$

with equality only when the solid is a sphere.
(b) Use the inequality to prove this: Of all solids of a given surface area, the sphere has the largest volume.

The three-dimensional **isoperimetric quotient** for a solid of volume V and surface area S is defined to be $36\pi V^2/S^3$. If the three-dimensional Isoperimetric theorem is true, then by problem 56 the isoperimetric quotient is always less than or equal to 1 with equality exactly when the solid is a sphere. The following table is more evidence for the truth of the theorem.

Solid	*Isoperimetric Quotient*
Sphere	1.0000
Icosahedron	0.8288
Best double cone	0.7698
Dodecahedron	0.7547
Best prism	0.6667
Octahedron	0.6045
Cube	0.5236
Best cone	0.5000
Tetrahedron	0.3023

From Pólya (1954), *Induction and Analogy in Mathematics.*

Problem 57

Verify some of the figures in the table above. (Try the cube, octahedron, and tetrahedron for a start.) Try to find a polyhedron with isoperimetric quotient higher than the icosahedron.

Problem 58

State and prove a three-dimensional analogue to Dido's problem.

Problem 59

The solution to Dido's problem is a consequence of the two-dimensional Isoperimetric theorem. Another consequence is that, of all polygons with sides a, b, c, . . . (in that order) the one inscribed in a circle has maximum area. Can you think of a three-dimensional analogue to this second consequence? Is it true? Can you prove it?

Problem 60

Think of three-dimensional analogues to other consequences of the two-dimensional Isoperimetric theorem. Are they true? Can you prove them?

Problem 61

[From Pólya (1954), *Induction and Analogy in Mathematics.*] You are thinking about the following problem: Find the polyhedron with a given number n of faces and given surface area that has the maximum volume. By analogy with two-dimensional problem (find the polygon with fixed number n of sides and fixed perimeter that has the maximum area), you suspect that, if there is a regular solid with n faces, then it is the one with maximum volume. However, it's wrong in two cases out of five! In fact the conjecture is

correct for $n = 4, 6, 12$

incorrect for $n = 8, 20$

What is the difference between the ones that are correct and the ones that aren't? (Observe some simple geometrical property that distinguishes between the two kinds of regular solids.) Why is the answer for $n = 8$ wrong? (Look at the preceding table for clues.)

Problem 62   The **convex hull** of a set S of points in the plane (or in space) is the smallest convex set containing the set S. If you have n points arranged on a circle, then the convex hull has maximal area when the points are equally spaced as the vertices of a regular n-gon. What about the three-dimensional problem? Of all arrangements of n points on a sphere, which one has the convex hull of maximum volume? If there is a regular solid with n vertices, then shouldn't it be the solution for that n? (Regular solids have $n = 4, 6, 8, 12,$ and 20 vertices.)

✳ *Full Circle*

A motivating theme of this chapter is to solve some geometric problems where you want to minimize this or maximize that. This is akin to the theme of Chapter 4 where we wanted to minimize path length. Another theme is one we've seen throughout the book: the use of analogy in solving problems. This theme occurs in many forms throughout the chapter: rectangle versus box, circle versus sphere, triangle versus tetrahedron. A third theme is analogy's companion, generalization; this occurs, for example, in moving from problems involving rectangles to the analogous problems involving pentagon, hexagon, and beyond, in moving from an area enclosed by a string, to an area enclosed by stick and string.

We proved two big theorems. The first was the theorem of the inequality relating the arithmetic and geometric means. Analogy and generalization played roles in the statement and proof of this theorem. The first version proved was for two variables, one of whose consequences resolved problems relating a rectangle's area to its perimeter. Considering analogous problems relating a rectangular box's volume to its surface area led us to generalize the theorem first to three variables, then to n variables (see problem 18). An interesting feature of the theorem is that it's basically a theorem about real numbers, even though we used it primarily to solve geometric problems. In the text it was mentioned that the full proof was incomplete, our proof being based on the assumption that a certain maximum *exists*. A proof without this assumption follows easily from Jensen's Inequality (see problem 7 of Chapter 8.)

The second big theorem was the Isoperimetric theorem. This theorem (about the perimeters of all regions having a fixed area) is itself a generalization of the chapter's first result (about the perimeters of all rectangular regions having a fixed area). Again, our proof was incomplete, being based on the assumption that a solution *exists*. This issue is discussed in this chapter's NOTES, where locations of complete proofs are pointed out.

✳ *Notes*

Babylonian scribes (ca. 2000 B.C.) made lists of problems of the form, given $x + y = A$ and $xy = B$, find x and y. Perhaps they did this to illustrate that rectangles of the same perimeter could have very different areas. [See Katz (1993), p. 32.] The fact that of all rectangles of given perimeter the square has greatest area is an easy consequence of Euclid's Proposition 5, Book II. The proof of this looks a lot like our geometric proof in this chapter. Pappus of Alexandria (early fourth century A.D.) displays the arithmetic and geometric means of two lengths in the following diagram (from Book III of his *Collection*).

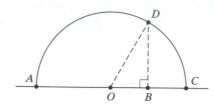

Lengths *DO* and *DB* are the arithmetic and geometric means, respectively, of *AB* and *BC*. From that he can easily conclude that *DO* is bigger than or equal to *DB*. As for the inequality involving more than two lengths, it appears that the Edinburgh mathematician Colin Maclaurin (1698–1746) was the first to state and prove it (1729) in the following geometrical form:

> If the Line *AB* is divided into any Number of Parts *AC*, *CD*, *DE*, *EB*, the Product of all those Parts multiplied into one another will be a Maximum when the Parts are equal among themselves.

Maclaurin's proof is close to the one you have seen in this book.

Since Dido's problem appears in the *Aeneid*—the epic by the Roman poet Virgil (contemporary of Augustus Caesar)—and since the *Aeneid* apparently **says** that the solution is a semicircle, the ancients probably knew something of the isoperimetric problem and its solution. Theon of Alexandria (A.D. late fourth century) and Pappus report on a book by Zenodorus (sometime between 200 B.C. and A.D. 100) in which he proved the following theorems:

- Among *n*-gons of the same perimeter, the regular one has the greatest area.
- Among regular polygons of equal perimeter, the one having more sides has the greater area.
- The circle has greater area than a regular polygon of the same perimeter.
- Of all solids with the same surface area the sphere has the greatest volume.

According to Kline (1972), "The subject of the theorems, . . . maxima and minima problems, was novel in Greek mathematics."

The proof we have given for the Isoperimetric theorem is due to the Swiss-born geometer, Jacob Steiner (1796–1863), in Berlin in 1836. Although Steiner thought he had proved the Isoperimetric theorem, Dirichlet pointed out that he had not proved the actual existence of a maximum, only that, if a curve exists enclosing a maximum area, it must be the circle. (The same gap occurred in Zenodoros's proof.) In the 1870s, Weierstrass of the University of Berlin filled in the missing piece. See Rademacher and Toeplitz (1957) for a discussion of the missing piece. The references listed below by Niven (1981), Courant and Robbins (1973), and Pólya (1954) have proofs not assuming existence.

✳ *References*

Courant, R., and H. Robbins, *What is mathematics? An elementary approach to ideas*. New York: Oxford University Press, 1973.

Fejes Tóth, L., *Regular figures*. New York: Macmillan, 1964.

Hildebrandt, S., and A. Tromba, *Mathematics and optimal form*. New York: Scientific American Library: Dist. by W. H. Freeman, 1985.

Katz, Victor, *A history of mathematics: an introduction*. New York: Harper Collins, 1993.

Kline, Morris, *Mathematical thought from ancient to modern times*. New York: Oxford, 1972.

Niven, Ivan, *Maxima and minima without calculus*. Washington D.C.: Mathematical Association of America, 1981.

Pólya, George, *Induction and analogy in mathematics, Vol. I of Mathematics and plausible reasoning*. Princeton, NJ: Princeton University Press, 1954.

Rademacher, Hans, and Otto Toeplitz, *The enjoyment of mathematics*. Princeton: Princeton University Press, 1957.

Steinhaus, Hugo, *Mathematical snapshots*. New York: Oxford University Press, 1950.

Stevens, Peter S., *Patterns in nature*. Boston: Little, Brown, 1974.

8 Beehives And Other Packing Problems

Take a large bunch of items of the same size and shape. You want to arrange them so that they take up the least amount of space. Since they may not fit neatly together, as in a tessellation, this means that we want to minimize the total amount of "wasted" space "between" items. In this chapter, we will consider problems of this kind, first in the plane, then later in space. Such optimization problems remind us of those we discussed in Chapter 4, where we minimized the lengths of paths, and in Chapter 7, where we minimized perimeters (given fixed areas) and minimized surface areas (given fixed volumes). We will keep these problems and their solutions in mind as we proceed. We will also encounter shapes (and arrangements) we met in other chapters. However, shapes, arrangements, problems and solutions from previous chapters come together in new and unusual ways in this chapter. This makes the trip exciting. This chapter is different in other ways. It's much more experimental and open than earlier ones: some problems we will look at are as yet unsolved in the mathematical world. This means that there will be more to do and more trips to take!

✳ The Irrigation Problem

You irrigate your alfalfa fields with pipes that rotate about a center. Thus, land where alfalfa is grown consists of a bunch of circles. Some of the land will not be used. How should the circles be placed in order to maximize the use of the land for growing alfalfa? Two possible configurations for the circles are shown below.

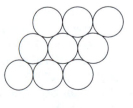

(Of course, we're thinking that the field is very large and very flat and consists of a lot of circles, a lot more than in these pictures.)

> **STOP! Try solving this problem before reading on.**

Getting Started with the Irrigation Problem.   To get a feeling for the problem, put a bunch of pennies on a table. Try to place them so that they do not overlap and so that they sit as close together as they can. You want the total amount of space between all the pennies—the "wasted space"—to be as small as possible. Some of the wasted space for the two configurations is shaded in the diagram below.

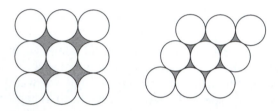

We can measure wasted space for a small piece of the field. But how can we measure it for a large field, one containing a large number of circles? This is not difficult for the configuration on the left extended in all directions indefinitely. Think of each circle sitting in a unique square, and all such squares tiling the plane. In one of these little squares (of side s) the wasted space is the area of the little square minus the area of the circle inscribed inside:

$$s^2 - \pi(s/2)^2 = s^2(1 - \pi/4)$$

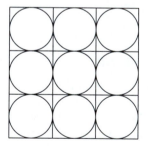

The fraction of the wasted space in the little square is $1 - \pi/4 \sim 0.2146$. This should be the fraction of wasted space for a lot of circles arranged in this way, sitting inside tiling squares. The fraction of useable space, $\pi/4 \sim 0.7854$, is also called the **density** of the **circle packing**. We want to find a packing of circles that maximizes this density.

Let's try to calculate the density of the other circle packing, the arrangement on the right extended in all directions. In this case you can think of each circle as inscribed in a unique regular hexagon, all such hexagons tiling the plane.

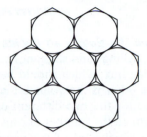

The fraction of useable space in a large bunch of such hexagons should be the same as the fraction of useable space in a single hexagon. The latter is the area of the inscribed circle (of radius r) divided by the area of the enclosing hexagon:

$$\pi r^2/(2\sqrt{3}\, r^2) = \pi/(2\sqrt{3}) \sim 0.9069$$

$[(2\sqrt{3})/3]r$

This is **much** better than the circles arranged in the squares. But is this the best? Wait and see. In the meantime, let's look at a related problem.

✻ *The Soup Can Packing Problem*

The "normal" way to pack cylindrical cans of soup in a packing box (or to pack eggs in an egg carton) looks like the following, from the top:

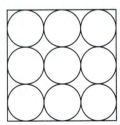

As with the alfalfa field, there is a certain amount of wasted space. This wasted space may cost money. Is there a less expensive way of packing cans?

!

STOP! Try this before reading on.

A Start on the Soup Can Packing Problem.   Let's think about the expenses involved in packing (and shipping) cans. Two items come to mind. One is the cost of the packing box, which should be roughly proportional to the surface area of the box. The other is the volume the box occupies in the truck. We can calculate both of these for the two different configurations we've been looking at. A third variable might be the number of cans to be packed in each box. (A fourth variable might be the dimensions of the array. For example, if we are to pack 16 to a box and the array is rectangular, should the array be 4 × 4 or should it be 2 × 8?) Let's assume for the moment that we will be packing 16 to a box in a 4 × 4 array. There are two obvious choices:

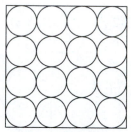
"square" arrangement

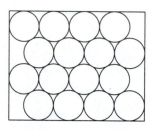
"hexagonal" arrangement

It's easy to calculate the two quantities for the choice on the left. Let h denote the height of the can and r its radius.

The volume of the box is then $(8r)(8r)(h) = 64r^2h$ and its surface area is 4 times the area of one side plus 2 times the area of top $= 4(8r)(h) + 2(8r)(8r) = 32rh + 128r^2$.

To calculate the two quantities for the choice on the right is a little trickier, but not too much so. Once we know the value of x in the diagram on the left below, the calculations will be easy as indicated in the diagram on the right.

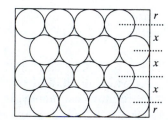

Thus, $x = \sqrt{3}\,r$ and the dimensions of the box are as in the diagram below.

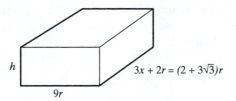

The volume of the box is $9r(2 + 3\sqrt{3})rh = 9(2 + 3\sqrt{3})r^2h \sim 64.77r^2h$. Its surface area is $2(2 + 3\sqrt{3})rh + 18rh + 18(2 + 3\sqrt{3})r^2 \sim 32.39rh + 129.53r^2$. It looks like the "normal" way ("square" packing) is better! (But not by much. . .)

Problem 1 The square packing of cans is better than hexagonal packing for a 4×4 array. Consider $n \times n$ arrays of cans for n large. Compare the square packing and the hexagonal packing of cans for such an n. What happens?

Problem 2 In our start at a solution to the Soup Can Packing Problem, we placed our 16 cans in a 4×4 array. Would some other arrangement be better, say a 2×8 array? In general, if you have m cans to pack, what's the most efficient array?

Problem 3 In our start to a solution of the Soup Can Packing Problem, we assumed that the box in which we pack the cans is a rectangle. For the hexagonal packing this means that there's lots of wasted space in the box. For the hexagonal packing let's replace the rectangular box by the parallelepiped below.

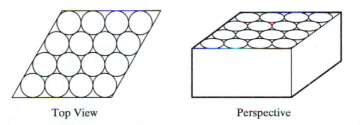

Top View   Perspective

Calculate the volume and the surface area of this new box. What happens? Is the size of the array an important factor here? (Compare with problem 1.) What are the advantages and disadvantages to building boxes in this way?

✴ The City Elementary School Partitioning Problem

The City has several elementary schools and has been trying to come up with a reasonable scheme to partition the City geographically so that the children living in each part will be assigned to a unique school. The School Board Partitioning Task Force has come up with the following criterion for determining each part:

For a given school, the part of the City assigned to that school should be those locations closer to the given school than any other school. As a member of the Task Force your job is to draw up the actual partitioning of the City satisfying this criterion. To aid you in this, a map of the City and its elementary schools is shown below.

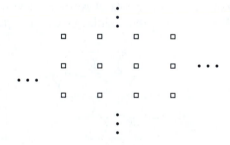

The part assigned to each school is called its **Dirichlet domain**.

Problem for right now!

Find the partitioning of the City into its Dirichlet domains. If you can, work on this with a group of two or three others.

Problem 4

Find the Dirichlet domains when the Elementary Schools are integer lattice points, as in the diagram below.

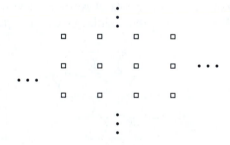

Problem 5

Find the Dirichlet domains when the Elementary Schools arranged in a triangular lattice, as in the diagram below.

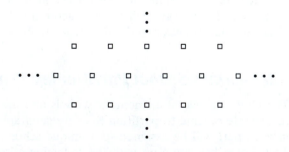

A Solution to the Irrigation Problem, Continued. We know that, for a large farm divided up into circular fields (the irrigated part) and the rest (the wasted part), the "hexagonal" arrangement is better (wastes less space) than the "square" arrangement.

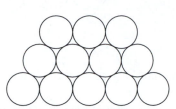

Hexagonal arrangement   Square arrangement

Could there be another arrangement that is better than the hexagonal one? What does it mean to be better? And how could you tell? (Of course, what we are looking for is the **best** arrangement.) Let's try to build the arrangement one circle at a time. Start with one circle.

Then add another. To avoid wasted space, you'd want the second to be tangent to the first.

Then add a third. Again, to avoid wasted space, you'd want the third circle to be placed in the "nest" between the first and second.

Keep doing this. Place additional circles in the nests between circles already there.

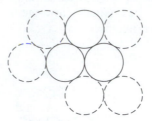

As you do this, you notice that the arrangement you get is the familiar "hexagonal" arrangement you've already seen.

It certainly feels like the hexagonal arrangement is the best. But how do you know, for some large collection of circles in a large region, that there isn't some other arrangement that is better? The next two sections provide an argument confirming our feeling that the hexagonal arrangement is always the best. Jensen's inequality will be used in the proof that the hexagonal packing of circles is best.

* Jensen's Inequality (Optional)

The Theorem of the Arithmetic and Geometric Mean says, if $x_1, \ldots, x_n$ are positive numbers then

$$x_1 \ldots x_n \leq [(x_1 + \cdots + x_n)/n]^n$$

with equality exactly when $x_1 = \cdots = x_n$.
If you take the log of both sides of this inequality, you would have

$$\log(x_1) + \cdots + \log(x_n) \leq n \log[(x_1 + \cdots + x_n)/n]$$

with equality exactly when $x_1 = \cdots = x_n$. Is there something special about the log function that makes this inequality true? Are there other functions $f(x)$ for which

$$f(x_1) + \cdots + f(x_n) \leq n f[(x_1 + \cdots + x_n)/n]? \tag{*}$$

Or, even simpler, how about functions for which

$$f(x_1) + f(x_2) \leq 2f[(x_1 + x_2)/2]? \tag{\#}$$

Here's what this says geometrically:

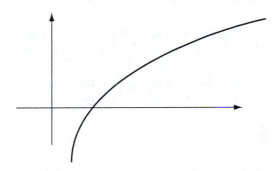

We'd want this to be true for all x's in a certain interval. It seems that the graph for f should be concave down in the interval from which the x's are to be chosen. Let's check this with the graph of the log function:

Sure enough, the log function is concave down for all positive x's. But the full inequality—for n numbers—is also true for logs. Is it possible that the full inequality (*) follows from truth for $n = 2$ [inequality (#)]? Let's try.

From several uses of (#),

$$f(x_1) + f(x_2) + f(x_3) + f(x_4) \leq 2f([x_1 + x_2]/2) + 2f([x_3 + x_4]/2)$$

$$= 2\{f([x_1 + x_2]/2) + f([x_3 + x_4]/2)\}$$

$$\leq 4f([x_1 + x_2 + x_3 + x_4]/4).$$

So the inequality is true for $n = 4$. A similar argument shows that the inequality is true for $n = 8, 16, \ldots$—any power of 2. If we could show that truth of (*) for m implies truth for $m - 1$, then we would know that (*) is true for all n. Let's try that, too.

Suppose (*) is true for m and that $x_1, \ldots, x_{m-1}$ are numbers in an interval where f is concave down. Let

$$x_m = (x_1 + \cdots + x_{m-1})/(m - 1)$$

Then,

$$f(x_1) + \cdots + f(x_{m-1}) + f(x_m) \leq m f[(x_1 + \cdots + x_{m-1} + x_m)/m] = mf(x_m)$$

Thus,

$$f(x_1) + \cdots + f(x_{m-1}) \le (m-1)f(x_m)$$

This is what we wanted to show. In conclusion, we have **Jensen's Inequality:**

If $x_1, \ldots, x_n$ *are in an interval where the function* f *is concave down, then* $f(x_1) + \cdots + f(x_n) \le n \, f[(x_1 + \cdots + x_n)/n]$ *with equality exactly when* $x_1 = \cdots = x_n.$

Problem 6

In our proof of Jensen's inequality we didn't worry about "equality exactly when. . ." Why is it valid to include this phrase?

Problem 7

Use Jensen's inequality to prove the Theorem of the Arithmetic and Geometric Means without assuming that a solution to the maximization problem exists.

Problem 8

There is another version of Jensen's inequality for functions whose graph is concave up. What is the inequality in this case? How does the proof go?

Problem 9

Jensen's inequality can provide us with an alternate proof to the theorem.

Of all n-gons inscribed in a given circle, the regular n-gon *has the greatest area.*

Here is a start. Consider the following pentagon inscribed in a circle.

Drop a perpendicular from the center of the circle to each of the sides and draw a line from the center to each vertex of the pentagon.

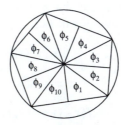

The area of the pentagon is $(1/2)r^2 \sum\limits_{i=1}^{10} \sin(\phi_i)\cos(\phi_i)$. This should look like something to which you can apply Jensen's inequality. Finish the argument for the pentagon. Then generalize. (Jensen's inequality thus makes a connection between the Theorem of the Arithmetic and Geometric Means and consequences of the Isoperimetric Theorem.)

Problem 10   Using an argument similar to that in the previous problem you can use the other Jensen's inequality to show that of all *n*-gons circumscribing a given circle, the regular circumscribing *n*-gon has the least area. (Can you think of a use for this result?)

✸ *A Proof That the Hexagonal Arrangement Is Best (Optional)*

Suppose we have a large, convex polygon *R* in the plane. Suppose there are a lot of our congruent, nonoverlapping circles in *R*. To keep things simple, let's suppose that each circle has area 1. (If our circles don't happen to have area 1, we could

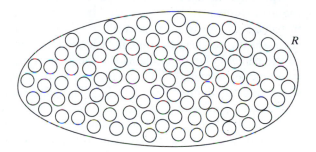

scale things so that they do.) What we want to do is compare the area of all the circles combined with the area of *R*. In fact, the **density** of the packing of the circles in region *R* is equal to the ratio of the area of all the circles to the area of *R*. This ratio is

$$\text{circle packing density} = n/A$$

where *n* is the number of circles in *R* and *A* is the area of *R*. Since we want this ratio to be as big as possible, it might be useful to obtain an estimate for it.

Now, for the square arrangement we considered earlier, making this estimate would be easy. For large convex *R*, the area of *R* would be roughly equal to the sum of the areas of the squares surrounding the circles.

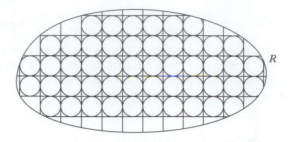

Thus the ratio above would be roughly

$$n/(n \times \text{area of single square}) = \text{area of circle/area of square} = \pi/4$$

which we got before.

Similarly, for the hexagonal arrangement, the area of R would be roughly equal to the sum of the little hexagons surrounding the circles and the circle packing density would be, roughly,

$$n/(n \times \text{area of single hexagon}) = \text{area of circle/area of hexagon} = \pi/(2\sqrt{3}).$$

This time, though, we have a bunch of circles inside R but we don't know what the arrangement is. However, let's try to place our cir les inside convex polygons in some way. We don't expect all the polygons to be the same. One way to do it is to use the solution to the City Elementary School Partitioning Problem: Think of the centers of each circle as the location of an elementary school and consider its Dirichlet domain. This domain is a convex polygon; the corresponding circle lies inside the domain, too. (Why?) The whole region R has been dissected into convex polygons with one, and only one, of our circles in each one.

Let's try to get some information about the polygons containing the circles. In particular, let's try to see if we can say something about the number of sides of the polygons. This sounds like a job for Euler's formula. Let s_i denote the number of sides of the polygon containing the ith circle. Let the data for the map be V vertices, E edges, and F faces. Then $V - E + F = 1$. If there are n circles, then $F = n$. On the boundary of R there are two types of vertices, those of order two and those of order three or more. Denote the number of the latter by T and M, respectively.

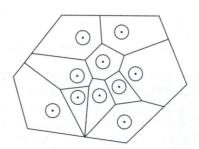

Then we have

$$6V - 6E + 6n = 6$$

$$3V \leq 2E + T$$

$$s_1 + s_2 + \cdots + s_n = 2E - T - M$$

Thus,

$$4E + 2T - 6E + 6n \geq 6V - 6E + 6n = 6$$

or

$$-2E + 2T + 6n \geq 6$$

or

$$2T + 6n - 6 \geq 2E$$

or

$$2T + 6n - 6 - T - M \geq 2E - T - M$$

Thus,

$$6n + T - 6 - M \geq s_1 + s_2 + \cdots + s_n$$

If R happens to be a hexagon, then $T \leq 6$ and

$$s_1 + s_2 + \cdots + s_n \leq 6n$$

i.e.,

$$\frac{s_1 + s_2 + \cdots + s_n}{n} \leq 6$$

In other words, the average number of sides to the polygons is less than or equal to 6. (That's amazing in itself!)

Now we're ready to start working on an estimate. Remember that we are trying to estimate n/A, where A is the area of R. As we've seen, it's useful to make R a (convex) hexagon. Let's go one step further and make it a regular hexagon. Denote by A_i the area of the polygon containing the ith circle. Then we have

$$A = A_1 + \cdots + A_n$$

What is the smallest that each A_i can be? First of all, A_i is bigger than (or equal to) the area A_i' of a polygon with the same number of sides that actually circumscribes the circle. Then, A_i' is bigger than (or equal to) the area $a(s_i)$ of the polygon with s_i sides that circumscribes the circle and has smallest area. What do you suppose the circumscribing polygon of smallest area is? The way things have been going in this book you would probably be inclined to say a regular s_i-gon. You would be right. (See problem 10.) Thus

$$a(s_i) = \text{area of regular } s_i\text{-gon circumscribing circle of area 1}$$

and

$$A \geq a(s_1) + \cdots + a(s_n)$$

Finally, we want to estimate the sum on the right. Let's see if we can get a formula for $a(m)$:

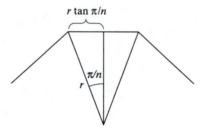

Thus, $a(m) = r^2 m \tan(\pi/m)$. Now $b(x) = x \tan(\pi/x)$ is a decreasing, concave (up) function.

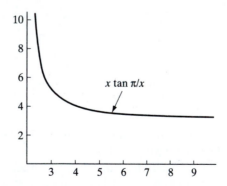

Therefore

$$b(s_1) + \cdots + b(s_n) \geq nb[(s_1 + \cdots + s_n)/n] \geq nb(6)$$

by Jensen's inequality. Consequently,

$$A \geq a(s_1) + \cdots + a(s_n) \geq r^2 n b(6) = na(6)$$

Thus,

$$n/A \leq n/[na(6)] = 1/a(6) = \pi/(2\sqrt{3})$$

Thus, the density can't be greater than the density for the hexagonal arrangement, no matter how large the region R. This proves that the hexagonal arrangement is the best.

✳ *Modeling Natural Phenomena with Arrangements of Circles*

The hexagonal arrangement of circles appears to be the best (meaning least amount of land wasted) for large alfalfa farms, irrigated in circular plots. The same should be true for packing cylindrical cans, provided the bottom of the rectangular box is "large" compared to the size of each little can. (See problem 1.) The hexagonal arrangement appears in nature in many different situations, but perhaps for different reasons. Here are a couple of experiments to try.

Experiment 1

Take a bunch of marbles (at least 25, all the same size) and place them in a shoe box lid. If you can, try to get them to arrange themselves in the square arrangement. You may want to start with the lid flat on the table. Then you'll want to tilt the lid at an acute angle to the table. Don't tilt too much or the marbles will spill out of the lid. What can you say about the eventual arrangement of the marbles?

Experiment 2

Take a bunch of tubes—such as paper towel tubes, toilet paper tubes—all the same size. Orient them all in the same direction and pile them all up on a table. (You will probably need to build some barriers at either end—out of books, say—so that you can build up several layers of tubes rather than have them all form a single layer line on the table.) What can you say about the eventual arrangement of the tubes? (Have a look, particularly at the arrangements of the circular ends of the tubes.)

For the marbles in the shoe box, you probably find it difficult to get them to stack in a square arrangement, especially after tilting the lid, even slightly, especially if there is any room at all for the marbles to move. The marbles persist in rolling into the valleys between their neighbors. In *Patterns in Nature,* Peter Stevens discusses this phenomenon further: "Marbles have less potential energy when they sit in the valleys rather than on the tops of other marbles; if marbles are free to move, they will automatically adopt a pattern of minimum energy." Minimum energy seems to be the ruling factor here rather than minimum wasted space; gravity rather than economy seems to be in charge. The same is true for the tubes: Gravity seems to force the tubes to arrange themselves in the hexagonal arrangement.

Marbles sitting in a shoe box lid model a layer of living cells—such as the cells of a layer of skin, of a leaf, or of the retina of a human eye. Typically, these cells grow uniformly and simultaneously. What happens? We assume that the cells are in a hexagonal array, based on the two experiments above. Each of the three cells around a "leftover space" expands equally to fill this space. The common tangent lines between pairs of circles form barriers beyond which no circle will expand. Moreover, the three tangent lines meet in a point and divide the "leftover space" into three equal pieces. Each circle then expands to fill one of these pieces and the three eventually meet at the intersection point of the tangent lines. This forms a tessellation of hexagons.

This is what may occur to cells undergoing internal pressure (growth). Examples of this in the real world are corn kernels, a cob of Zamia skinneri, yeast rolls, domains of coral polyps, and cells in a wasp nest.

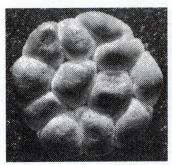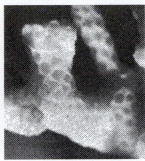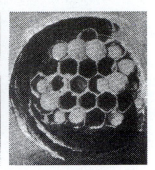

Instead of growing, the whole hexagonal arrangement of spheres or cylinders could be squeezed together. We imagine that each individual sphere is made of deformable material.

Experiment 3

Make a bunch of equal spheres or thick circular discs out of Play-Doh, plasticene, or modeling clay. (Use a round cookie cutter or doughnut form.) Arrange them so that they touch in the "hexagonal" arrangement. Color objects that touch with different colors or roll separate objects in chalk dust or flour. (The idea is to be able to identify separate objects in the end after all the compression has taken place.) Place the planar arrangement between two boards (or pieces of masonite or plexiglas . . .) and press the two together. What happens?

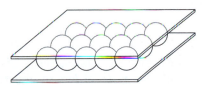

Similar forces or pressures seem to operate in creating the beehive. Bees make cylindrical cells which, because of gravity, get stacked in the hexagonal arrangement. Then "imagine the whole system under some uniform stress—of pressure caused by growth or expansion with the cells, or due to some uniformly applied constricting pressure from without. . . . the six points of contact between circles will be extended into lines, representing surfaces of contact. The equal circles are converted into regular and congruent hexagons." [*On Growth and Form,* d'Arcy Wentworth Thompson (1961), pp. 500ff.]

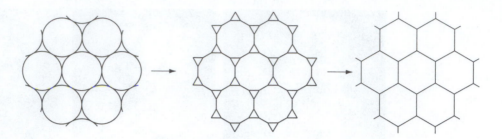

Problem 11

An electrical engineer packs wires around a central wire in the hexagonal arrangement. She wants to know the total number of wires used. Of course, this will depend on the number of layers (or "zones") built up around the central wire. Help her find the total number of wires given that the number of layers is n.

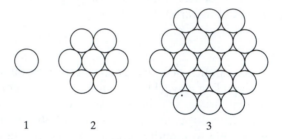

1   2   3

In the beehive and the pigmented epithelium of the retina, tessellations of perfect, regular hexagons occur. Many other tilings occur in both the artificial and natural worlds that come about for other reasons. The following experiment results in one of these.

Experiment 4 (with soap bubbles)

Between two plates of plexiglass, blow one bubble, then another next to it, then another next to them. Blow a bunch of these bubbles. You want them to cluster together and be more or less of equal size. Observe the pattern of tiling that occurs. What do you see? (What kinds of junctures occur? What kinds of tiles are there? How many sides to each polygonal tile?)

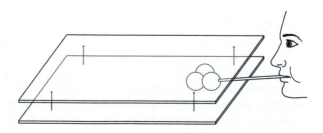

Here are pictures of the cells of a soap froth, the section of polycrystalline MgO, and the cell structure of the skin of a cucumber.

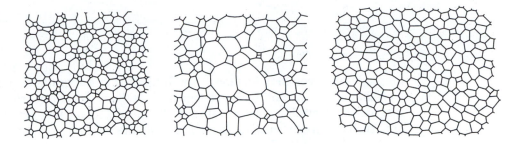

In *Connections* (pp. 215ff), author Jay Kappraff (1991) asks, "How is it that physical and biological systems that are influenced by such different external forces, nevertheless end up with similar patterns?"

In the examples of the Dirichlet regions for the school serving areas in a community considered earlier, we wind up with a geometric pattern interchangeable with

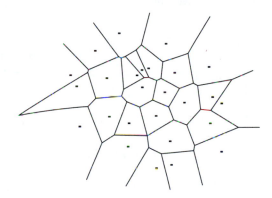

that of the soap bubbles, the animal tissue, and the plant tissue. One common feature to these patterns is the prevalence of vertices of order three. For example, you notice in the soap bubble configuration that all the vertices are of order three and that the angles are all 120°. This is a consequence of the solution to the Three Cities Problem and to the solution of problems 28 and 29 of Chapter 4 and its extensions in Chapter 9. Vertices of order three seem to be prevalent in nature— they are simpler and more stable. By the latter I mean that it is very difficult to create a vertex of order four or more: A minuscule perturbation of the ingredients will create two vertices of order three instead. But

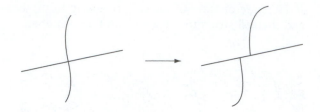

there are other common features. "Smith (quoted in Kappraff, loc. cit., pp. 215ff) feels that at the scale of these phenomena, it is the geometric constraints on space that are the controlling factors rather than external forces that determines form." One such constraint on space is the following theorem:

Theorem.   Suppose the plane is tiled with polygonal tiles such that all vertices are of order 3 and there is an upper and a lower bound on the size of each tile. Then the average number of edges per polygonal face is exactly 6.

Proof.   Consider a piece of the tiling. Think of the piece as an island and the sea as a separate face. This yields a decomposition of the sphere into vertices, edges, and faces. Denote by V, E, and F the corresponding data for this decomposition. Then we know that

$$V - E + F = 2$$
$$3V = 2E$$

and

$$2E = 2F_2 + 3F_3 + \cdots$$

where F_i = the number of polygonal faces with exactly i edges. Then

$$12 = 6V - 6E + 6F = -2E + 6F$$

Thus,

$$6F = 12 + 2E = 12 + 2F_2 + 3F_3 + \cdots$$

and

$$6 = 12/F + (2F_2 + 3F_3 + \cdots)/F$$

As the piece of the tiling gets bigger, the size of F gets bigger and the quantity $(2F_2 + 3F_3 + \cdots)/F$—the average number of edges per polygon—gets closer and closer to 6. This completes the proof.   ✳

Problem 12   How does this theorem jibe with what we know about polyhedra in terms of the average number of edges per polygonal face?

✳ *The Wine Bottle Packing Problem*

As with soup, the traditional way of packing wine bottles in a box is as follows:

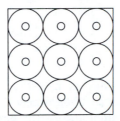

Unlike soup, wine is sold in glass bottles and, to protect the bottles from breakage, cardboard is placed between the bottles, like this:

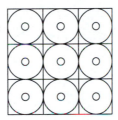

Of course this adds to the cost of packing. Is there another way to pack wine bottles so that this new expense is minimized? What is the best way?

STOP! Think about this before reading on.

A Start on the Wine Bottle Packing Problem.   As we did with the soup can packing problem, let's look at a simple case. Let's assume we are shipping 16 bottles of wine in a case. For us there are two favorite ways of shipping:

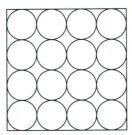

Square arrangement

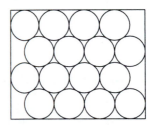

Hexagonal arrangement

Let's calculate how much extra cardboard is needed for each of these. The extra cardboard for the square arrangement is shown below.

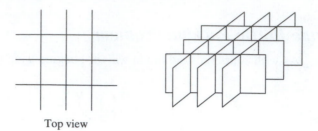

Top view

If the bottom (maximum) radius of each bottle is r and its height is h, then the amount of extra cardboard needed is $48rh$.

The extra cardboard for the hexagonal arrangement is shown below.

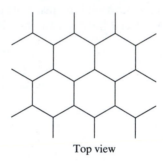

Top view

This is a bit more complicated. There are 12 (4 rows of 3 each) vertical lengths and 21 (3 rows of 7 each) horizontal "zig-zag" lengths, each length being the side of a regular hexagon inscribing the circular bottom of a bottle.

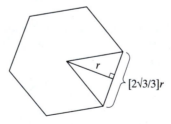

This length is $[2\sqrt{3}/3]r$. Thus, the total amount of extra cardboard needed is

$$33 \times [2\sqrt{3}/3]rh \sim 38.12rh$$

a lot less than for the square arrangement!

Problem 13   Calculate the amount of extra cardboard for an $n \times n$ array in two ways: (1) in case the bottles are in the square arrangement and (2) in case the bottles are in the hexagonal arrangement. What is the amount of extra cardboard needed per bottle in each case? (This last quantity should depend on n.) What happens to these two quantities as n gets large?

Problem 14   Based on the results of the last problem, the Soup Can Packing Problem, and of problem 1, how should you pack glass bottles in order to minimize costs? There are at least three cost factors to include in your discussion: cost of the materials used in making the big box, cost of the extra cardboard, and cost of the space the big box occupies. There may be other factors you want to discuss such as practical limitations on the size of the big box—if it's too heavy, then . . . ; if it's too big, it may be too awkward to handle . . . and so on.

If we want to explain why the beehive is the way it is, we might try to show that they build it that way because it uses the least amount of wax. So, suppose we were going to build a container (to hold honey)—not the usual box, but one assembled, as the beehive is, out of equal prisms whose ends tile the plane. If the container is made up of a lot of these prisms, then minimizing the ratio of amount of wax used to the volume of the container is the same as minimizing the ratio of wax in a single prism to the volume of the prism. The latter is the same as minimizing the perimeter of the tile to the area of the tile. Thus, of all polygons of fixed area we want to find one that

- Tiles the plane
- Has minimum perimeter

You will recognize this as an isoperimetric problem for tessellating polygons. We know several polygons that tessellate the plane: any triangle, any quadrilateral, a regular hexagon. It turns out that there are nonregular pentagons and hexagons that tile the

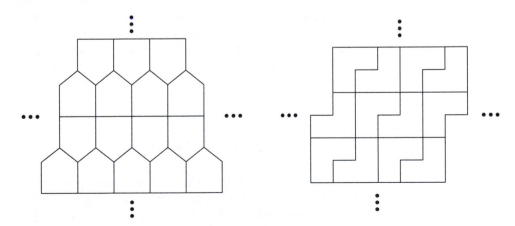

plane. We also know that of all triangles, quadrilaterals, pentagons, and hexagons of fixed area the regular hexagon has least perimeter. And we know that no regular *n*-gon for $n > 6$ tiles the plane. Could there be a nonregular 7-gon or 8-gon or . . . that tiles the plane with perimeter less than that of a regular hexagon of the same area? The answer is no. [See Niven (1981) listed in this chapter's references for a really interesting argument for this.] If the bees have economical foresight, then their solution is right on.

Problem 15   What does this tell you about a solution to the Wine Bottle Packing Problem?

✳ *The Ball Bearing Shipping Problem*

You are a manufacturer of ball bearings in Los Angeles and have been asked to make an emergency shipment of ball bearings, all the same size, to Honolulu. You are to ship as many as you can within 24 hours. Although you have a large stock of the right size ball bearings, you have been able to locate only one ship that can make a shipment on such short order. You tell the shipper that you want to ship as many ball bearings as you can. She asks for, and you give her, the weight and volume of a single ball bearing. Based on this information, she concludes that the draft of the ship would be too deep for the local channel if the ball bearings (not including the spaces between them) were to occupy more than three fourths of the volume of the hold. In order to ship as much as you can you would like to fill the hold completely (spaces included). Can you do this and still satisfy the restriction? (Thanks to N. J. A. Sloane.)

!

STOP! Think about this before reading on.

Getting Started on the Ball Bearing Shipping Problem.   You think: If the ball bearings were cubes and the hold rectangular, you could stack the cubes with no space between them. This way, you could only fill up three quarters of the hold. Of course, if you threw the cubes in willy-nilly, there would be more wasted space. But you will be safe no matter how you place them as long as you fill no more than three quarters of the hold. The ball bearings being spheres are a different story. No matter how you place them in the hold there will always be wasted space. What's the least amount of wasted space you could achieve with spheres? If this least amount is 25% or more of the volume, then no matter how you placed the spheres you could fill up the hold and the restriction would be satisfied. What we need to do now is figure out the best way (or ways) to pack spheres and determine the percentage of wasted space for this best way.

Problem 16   The size of the spheres (ball bearings) hasn't come up. Wouldn't the percentage of wasted space be greater if the spheres were larger? Compare this question with the analogous one in the plane: Doesn't the percentage of space wasted depend on the size of the circles?

✳ *Sphere Packing Experiments*

Here are some experiments that will give us a feel for packing spheres and, hopefully, help us solve the Ball Bearing Packing Problem.

Experiment 5

Get a bunch of styrofoam spheres (around 3-in diameter) and some toothpicks. Arrange the spheres as close together as you can. Connect adjacent spheres by toothpicks. Make several layers. Create some interesting designs. What can you say about the best way (or ways) to pack spheres? Best means, of course, that the amount of wasted space is minimal.

Experiment 6 (using analogy and what we know in the plane)

The analogous problem of closely packing circles in the plane suggests an approach to the problem of packing spheres in space. We could build up a packing of spheres layer by layer. In each layer we would want to pack the spheres as close together as possible by placing them in the familiar hexagonal arrangement.

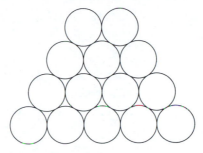

Each of the layers would look just like this first one. To see how the layers might stack, use styrofoam balls and toothpicks as before to make three copies of this layer. Start with two of these layers. Take two of these layers and place one on top of the other to see if they fit. Do they? Try fitting them in different ways. What do you observe?

Fit two layers together and add a third one. As before, try fitting the third layer in different ways. What do you observe?

Here are some diagrams to help see what is going on with Experiment 6. Here is the first layer.

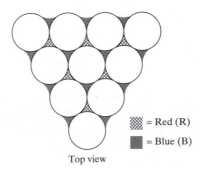

▨ = Red (R)

▦ = Blue (B)

Top view

Perspective

From the top you see the nests between the spheres of the first layer. We have colored them blue and red: Six blue and red nests alternate around each sphere. The spheres of the second layer fit in these nests. The second layer of spheres can't fill all the nests. In fact there is a choice. The second layer can fill all the red nests or it can fill all the blue nests.

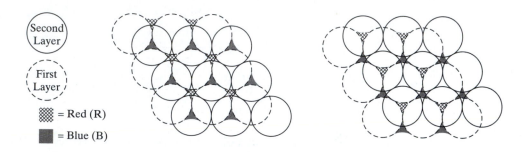

Second
Layer

First
Layer

▨ = Red (R)

■ = Blue (B)

Although there are two choices, the results for the two layers are essentially the same: One is a rotation of the other through 60° about an axis through the center of an original layer's sphere and perpendicular to the planes of the layers.

For the third layer there are also two choices. If the second layer fills the red nests, then the spheres of the third layer will either sit directly above the spheres of the original layer or above the blue nests of the original layer. We can denote these two possibilities by ORO and ORB, respectively. Similarly, if the second layer fills the blue nests, then the spheres of the third layer will either sit directly above the spheres of the original layer or above the red nests of the original layer. These two possibilities are denoted by ORO and ORB. The configurations OBO and ORO are essentially the same. So are ORB and OBR.

Experiment 6, continued

With styrofoam balls and toothpicks, construct ORO and ORB and compare. With styrofoam balls and toothpicks, construct OBO and ORO and compare. With styrofoam balls and toothpicks, construct OBO and OBR and compare. What are your conclusions?

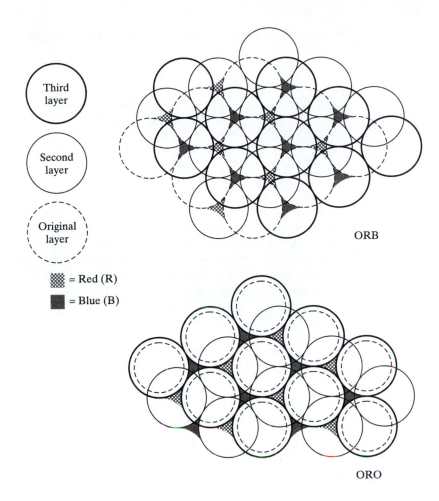

From the diagrams just above, it certainly seems that ORO and ORB are different. On the one hand, ORO has spheres centering above just two locations: the Os and the Rs. On the other hand, ORB has spheres centering above the three locations: the Os, Bs, and Rs. In fact, in ORO you can **see** through the B locations from top to bottom, whereas in ORB there are no openings to see through from top to bottom. Let's see if we can find another way to distinguish the two.

Experiment 6, concluded

Take out your styrofoam-toothpick ORO and ORB. For each one, select a sphere in the middle layer completely surrounded by other spheres. How many other spheres does it touch? For each case, try to figure out what the configuration— single sphere plus all the spheres touching it—looks like.

In the middle layer of each one, there should be six spheres that touch the selected one.

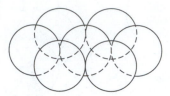

In the top layer there are three additional spheres that touch the selected one—similarly for the bottom layer. Thus, six on the same layer, three on the layer above, and three on the layer below: 12 spheres in all. It doesn't matter whether it's the ORO or the ORB configuration: 12. Interesting, but no help yet in telling them apart.

Now let's have a look at these 12 spheres. For the ORO and the ORB configurations, the nine spheres of the bottom and middle layer look the same:

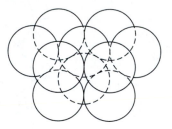

Adding the three spheres of the top layer, we get the following:

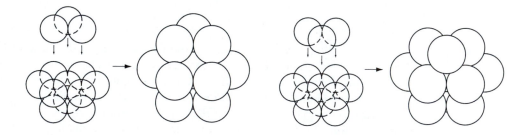

In the ORB case, the centers of the 12 spheres form the vertices of a cuboctahedron, what you get when you slice off the vertices of a cube leaving a polyhedron with faces consisting of squares and equilateral triangles.

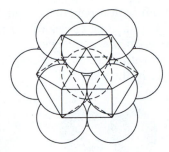

In the ORO case, the centers of the 12 spheres form the vertices of a polyhedron gotten by twisting a "cap" of the cuboctahedron 45°.

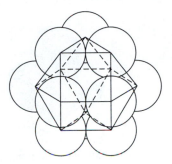

The two polyhedra are clearly different.

Continuing to add layers in this manner, we could get the two sequences of layers OROROR ORO . . . and ORBORBORB . . . The configuration of a single sphere and its touching neighbors in such packings of spheres should be a twisted cuboctahedron and a cuboctahedron, respectively.

Problem 17   OROROR ORO . . . and ORBORBORB . . . are not the only possibilities you get if you closely stack layers of close-packed spheres. What are the other possibilities?

The best way to pack circles in the plane also turned out to be the same as making every three neighboring circles pack as closely as they can in a triangular formation.

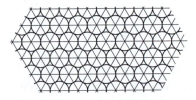

Thus, if we started out with a pair of tangent circles, added another to a "nest" between them, and kept adding circles to existing nests, then we would have the best packing of circles in the plane. Perhaps the same strategy would work for spheres in space. Start with a pair of tangent spheres. Add another in a nest between them to get a triangle in the plane.

Then, to get out of the plane of the triangle, add another sphere in the nest produced by the three triangles to get a tetrahedron.

The next step would be to add spheres to the nests produced by the spheres making up the tetrahedron.

Experiment 7   Carry out this plan with styrofoam balls and toothpicks. Make the tetrahedron. Fill in the nests and keep going. What happens?

Experiment 8   Equilateral triangles tessellate the plane. The shape in space analogous to an equilateral triangle is a regular tetrahedron. Take a bunch of congruent regular tetrahedra. Try to arrange them in space to make a space filling tessellation. What happens? What does this have to do with Experiment 7?

Problem 18   Consider the arrangement of four tangent spheres whose centers form the vertices of a regular tetrahedron. Calculate the volume of the whole tetrahedron. Calculate the volume of that part of the tetrahedron occupied by pieces of the packed spheres. Suppose the strategy for packing spheres suggested by Experiment 7 worked so that space could be tessellated with regular tetrahedra. What percentage of the volume would be occupied by spheres and what percentage by empty space?

Problem 19   Was the strategy for stacking spheres suggested by Experiment 7 followed in creating the sphere layers OROROR . . . and ORBORBORB . . . ?

Our two favorite arrangements of circles in the plane are the square arrangement and the hexagonal arrangement. We utilized the hexagonal arrangement in creating the layer sequences of spheres OROROR . . . and ORBORBORB . . . . This time let's use the square arrangement. Of course, the square arrangement was not optimal in the plane, so we haven't got high hopes for it in space. Working with it may give us some ideas.

Experiment 9   With styrofoam balls and toothpicks, make a couple of layers of spheres in the
(An analogy   square arrangement.
with the plane
that probably
won't work)

Experiment with ways to stack the layers. Try to find ways that yield the least amount of wasted space. What do you observe?

Problem 20   One way to stack layers of spheres in the square arrangement is to place the layers so that the centers of spheres are directly on top of one another:

This is the way oranges may be packed in a box for shipping. Calculate the percentage of wasted space in this stacking of spheres.

Travelling around the country you can still see, on county courthouse lawns, old war memorials made of cannon balls stacked in the shape of Egyptian pyramids.

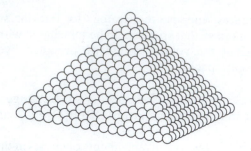

Experiment 9, continued

Take a layer of styrofoam balls and toothpicks arranged as a 4 × 4 square. Using more styrofoam balls, construct an Egyptian pyramid on top of this, i.e., the second layer should be a 3 × 3 square array of spheres, and so on. How do successive layers fit? Are there choices as there were in making ORORO . . . . and ORBORB . . . ? Ponder the completed pyramid. In particular, look at the slanted sides of the pyramid. Any observations?

You may have observed that the spheres of a slanted side of the cannonball pyramid are arranged in a hexagonal arrangement. What a surprise! We stack square arrangements of spheres and a hexagonal arrangement pops up!

Experiment 9, concluded

Turn your cannonball pyramid over and sit it on one of these slanted sides. You can now think of this pyramid as made up of layers of spheres parallel to this new base. Each layer is a hexagonal arrangement of spheres in a plane. Can you say anything about how these layers are stacked?

The slanted side layers of the cannonball pyramid are stacked as the layers in OBROBROBR . . . ! Another real surprise! It seems as if the inefficiency (and instability) of the square arrangement is compensated for by the deeper nests of a square arrangement as compared to the nests of a hexagonal arrangement.

Problem 21

Calculate the depth of the nests for the cannonball layers (bottom layer with spheres in square arrangement) and the depth of the nests for the ORBORB . . . layers. Compare.

Problem 22

Find the number of cannon balls in an *n*-layer square-based pyramid. (Compare with problem 24 of Chapter 1.) Estimate the volume of the underlying pyramid approximated by the cannon balls, including the wasted space. Estimate, for large *n*, the percentage of wasted space between cannon balls.

Problem 23

Create a triangle-based pyramid out of spheres. The spheres forming the base should be arranged as an equilateral triangle of side *n* in the hexagonal arrangement. The next layer should be an equilateral triangle of side *n* − 1, and so on. How do the layers stack? (ORORORO . . . ? ORBORBORB . . . ?) You might want to make a model of this out of styrofoam balls and toothpicks. As in the previous problem, find the number of spheres used to make this n-layered pyramid. (Compare with problem 23 of Chapter 1.) Estimate the volume of the underlying pyramid approximated by the cannon balls, including the wasted space. Estimate, for large n, the percentage of wasted space between cannon balls. How does this percentage compare with the percentage calculated in the previous problem?

✳ *A Return to the Ball Bearing Problem*

It turns out that, when this book went to press, ORBORBORB . . . wastes less space than any other known packing of spheres. The arrangement is often called **cubic close packing.** At that time there was no proof that a packing wasting even less space did not exist. Of course, OROROR . . . and other variations of OROROR . . . and ORBORBORB . . . all waste the same amount of space as ORBORBORB . . . . (Why is this?) The current conjecture is that the percentage of wasted space for ORBORBORB . . . is least. Thus, to get the best thinking to date on the Ball Bearing Problem we should calculate what the fraction of wasted space is or, equivalently, the fraction of nonwasted space. The latter number is known as the **density** of the packing. Actually, the results of problems 22 and 23 should enable us to calculate this number. Let's try another method suggested by what we did in the plane. When we calculated the density of the hexagonal arrangement of circles in the plane, we noted that each circle was circumscribed by a unique regular hexagon and that all the hexagons tiled the plane. In fact the hexagons were the Dirichlet regions corresponding to the centers of all the circles. (See problem 5.) The density of the circle packing is the ratio of the area of the circle to the area of its enclosing polygon. If this approach works in space, then for every sphere in the cubic close packing, we ought to be able to find a unique polyhedron that circumscribes each sphere so that all the polyhedra tile space. Then, the density of the cubic close packing would be the ratio of the volume of a sphere to the volume of a polyhedron.

To carry out this plan, it would be useful to look at the cubic close packing in another way. To do this, consider a filling of space by cubes. Color the cubes white and black, checkerboard style: Cubes that meet along a common face should be colored different colors.

Throw away all the white cubes and place spheres at the center of each black cube so that each sphere passes through the midpoint of all its cube's edges. (Try making this arrangement of spheres out of styrofoam balls and toothpicks.)

Each checkerboard layer looks like the spheres packed in the square arrangement. These layers stack like the layers of the cannonball pyramid. Thus, this is another description of the spheres in cubic close packing!

Next, dissect each of the white cubes into six pyramids which meet at the center of the cube. (At this point you'll want to take out the model you made for problem 32 of Chapter 2.)

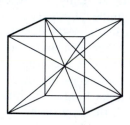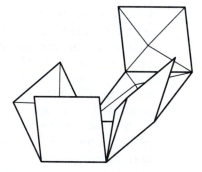

Finally, glue the base of each white pyramid onto the face of the black cube it abuts. Gluing white pyramids onto all the faces of each black cube turns each black cube into a rhombic dodecahedron circumscribing the sphere whose center is at the center of the cube.

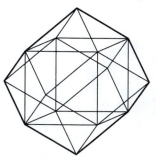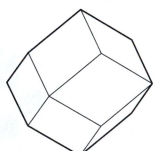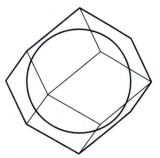

(Another way to describe this construction would be to turn each black cube, dissected into **its** pyramids, inside out. Check this out with the model from Chapter 2 just mentioned.)

Thus, every sphere of the cubic close packing sits inside a unique rhombic dodecahedron. Moreover, from our construction you can tell that the rhombic dodecahedra tile space (i.e., every spot is contained in one of the polyhedra and there are no overlaps). Our strategy is successful!

Problem 24   Check that each rhombic dodecahedron really is the Dirichlet region of the center of the sphere it contains. [Hint: Select a sphere and a face of the rhombic dodecahedron circumscribing it. An adjacent sphere lies on the other side of that face. Take the line segment joining two centers of these two spheres and consider the perpendicular bisector of that segment. (It's a plane!) What's the relationship between the plane and the face?]

Problem 25   The following is a pattern for a rhombic dodecahedron. Make a bunch of copies of this pattern. Then cut out each one, fold on solid lines (away from you), and assemble with glue on the tabs or with tape. Using a bunch of the assembled polyhedra, show that they tessellate space. You might want to try doing this using The Geometry Files.

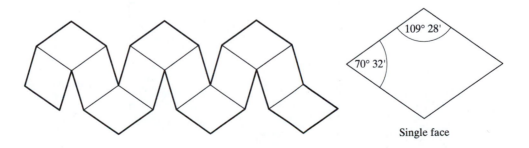

Single face

To calculate the density of the cubic close packing, we need to find the ratio of the volume of a sphere to the volume of the rhombic dodecahedron enclosing it. If each cube has side one, then the volume of the cube is 1 and the volume of the rhombic dodecahedron is 2. (Why?) The sphere has radius $\sqrt{2}/2$ and volume $(4/3)\pi(\sqrt{2}/2)^3 = \pi\sqrt{2}/3$. The ratio of volumes is $\pi\sqrt{2}/6 \sim 0.74$. If the sphere packing conjecture is correct, .74 is the largest possible density for sphere packing. Thus, if the hold of the ship is filled up completely with ball bearings (even in a willy-nilly fashion), at most .74 of the volume of the hold will be made up of solid ball bearings.

Problem 26   How does the density just calculated compare with the estimates for the density obtained in problems 22 and 23?

Experiment 10   Rhombic dodecahedra are what you imagine getting if you subject all the spheres in a cubic close packing to some kind of internal pressure (for example, from growth—a pressure acting uniformly outwards from the center of each sphere) or external pressure (acting inwards from all directions equally). Take a bunch of spheres made of some compressible material, such as soft styrofoam. Arrange them in cubic close packing. Subject them to pressure, uniform from all directions. Observe the shapes the spheres turn into. What do you see?

✳ Caps for the Beehive Cells

Earlier in the chapter, we discussed the hexagonal form taken on by the opening of a bee cell. Then, we imagined that the cells had started out being cylindrical with the circular openings in the hexagonal arrangement. Whatever forces the whole conglomerate was subjected to—internal growth from within or pressure from without—the hexagonal arrangement of circles turned into a tessellation by regular hexagons.

There is more to a beehive than the tessellation of hexagons formed by the cell openings. Each cell has a closed end as well. Moreover, two families of cells meet along these closed ends. At each cell end is part of a rhombic dodecahedron. If you look at a rhombic dodecahedron head on with one of its order three vertices in the middle, you will see the outline of a hexagon and understand how a piece of a rhombic dodecahedron can fit nicely on the end of a hexagonal prism.

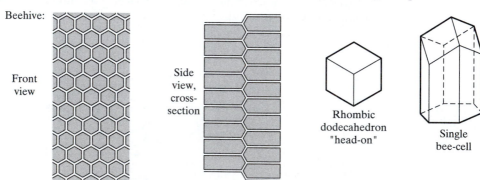

Beehive:

Front view

Side view, cross-section

Rhombic dodecahedron "head-on"

Single bee-cell

But how do those rhombic dodecahedra end up there? Why not some other shape? Here is an explanation.

Suppose as before that each bee cell begins as a cylinder. This time, however, it has to have a closed end. What should the closed end look like? A cylinder capped with a hemisphere would minimize the surface area for given volume. (Check this out!) So imagine a bunch of cylindrical cells with hemispherical ends stacked in a hexagonal arrangement. Imagine an identical configuration of cells back-to-back against the first. How should the two sets of sphere-ends fit together? Yes: cubical close packing. (Of course, we're only talking about two layers of cubical close packing.)

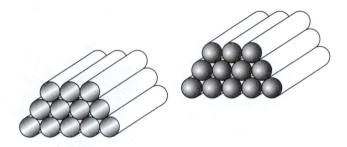

Now put the whole system—the two sets of cells back-to-back—under some kind of pressure. The circular openings of the cylinders turn into regular, tessellating hexagons, and the hemispheres turn into pieces of rhombic dodecahedra that mesh together with no empty spaces.

We know that the regular hexagonal ends of the prisms result in the least amount of material for all tessellating prisms of fixed volume. Do the rhombic dodecahedral ends use the least amount of material for the job that they do? Wait and see. We'll deal with this issue soon.

✳ *Body Centered Cubic Packing*

We have discussed in detail three ways to pack spheres. There is the layering OROROR . . . , called **hexagonal close packing** (not to be confused with the hexagonal arrangement of circles in the plane). There is the layering ORBORBORB . . . called **cubic close packing.** For both of these, each sphere touches 12 others. There is **simple cubic packing** in which the centers of the spheres sit at integer lattice points in space. Each sphere in this packing touches six others.

A fourth way of packing spheres is to place spheres at integer lattice points and at the center of each of the cubes formed by the space grid. You want the spheres in this arrangement to be just large enough that the spheres at the cube's vertices are tangent to the sphere in the middle. This is called **body centered cubic packing.** Each sphere in this packing touches eight others.

Simple
cubic packing

Body centered
cubic packing

Problem 27

Calculate the density of this packing. (Hint: One way to do this might be to look at a cube whose vertices are eight nearby lattice points. The volume of that cube is 1. Then figure out how many spheres or fractions of spheres occupy the cube.)

For the cubic close packing, each sphere sits inside a unique rhombic dodecahedron. What is the analogous shape for body centered cubic packing? We could use the shape to find an alternative method for calculating the density of body centered cubic packing. First, an experiment.

Experiment 11

(1) Make two copies of the pattern below, cut them out, fold on solid lines (away from you), and assemble using glue or tape. The two assembled shapes should fit together to make a cube. Check it out!

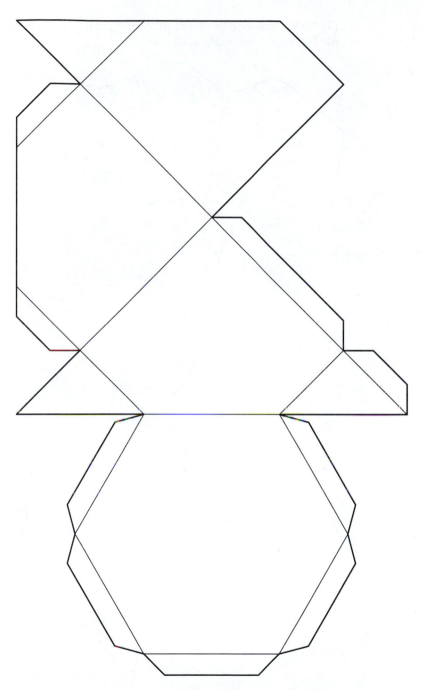

(2) Make six copies of the preceding pattern. Cut them out, fold, and assemble. These six shapes plus the two just made will fit together to make a truncated octahedron. Do this!

(3) Make nine copies of the following pattern for a truncated octahedron. Cut out, fold, and assemble each one.

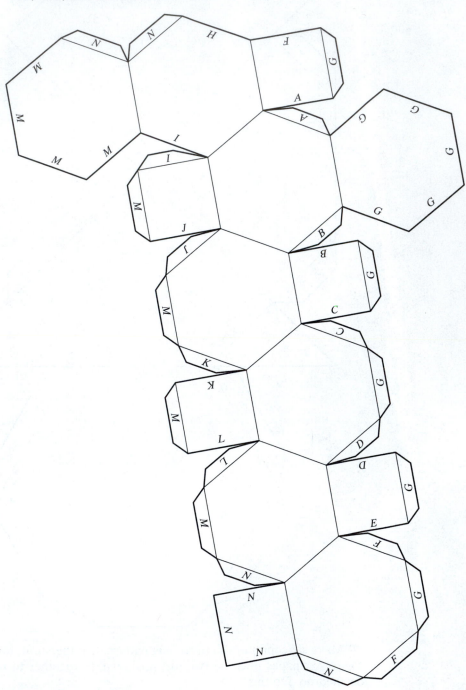

Play around with the completed polyhedra to see if they will tile space. Show that the nine fit together to make the following configuration. Alternatively, you might want to carry out (3) using The Geometry Files.

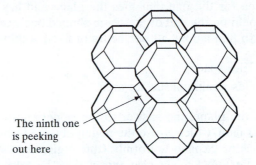

The ninth one
is peeking
out here

Let's return to body centered cubic packing. We claim that each sphere in body centered cubic packing is enclosed in a unique truncated octahedron and that all the truncated octahedra fill space without gaps or overlaps. The following exploded views of body centered cubic packing demonstrate this.

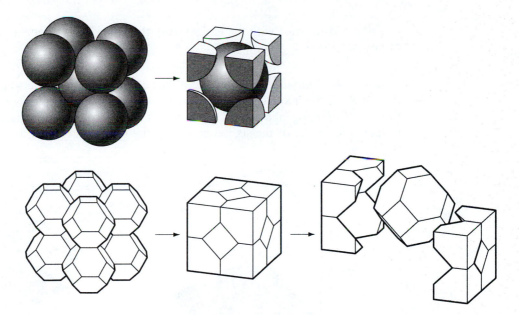

✵ *Beehives and the Isoperimetric Problem for Space-Filling Polyhedra*

In an early discussion of bee cells we were looking among all polygons of fixed area for the one that tiles the plane and has minimum perimeter. (The polygon would be the end of the prism-shaped bee's cell.) The analogous problem in space is to find among all polyhedra of fixed volume the one that

- Tiles space
- Has minimum surface area

In the plane, the answer to the isoperimetric plane tiling problem is the regular hexagon, which is also the region obtained when all the circles in the densest plane packing expand uniformly (this region is also the Dirichlet region for a circle's center). We might also expect that the solution to the isoperimetric space tiling problem would be the same as the region obtained when the spheres in the densest known packing in space expand uniformly (this region is also the Dirichlet region for a sphere's center). The latter is the rhombic dodecahedron. Let's compare it to two other space filling polyhedra that we know: the cube and the truncated octahedron. We will assume that each polyhedron has volume 1.

Cube. Let's start with the easiest case, the cube. The surface area of the cube of volume 1 is 6.

Rhombic Dodecahedron. The volume of the rhombic dodecahedron is twice that of the cube it surrounds. So the cube has volume 1/2 and the cube's edge is thus $1/\sqrt[3]{2}$. The surface area of the rhombic dodecahedron is 24 times the area of a slant side of one of the six pyramids making up the polyhedron.

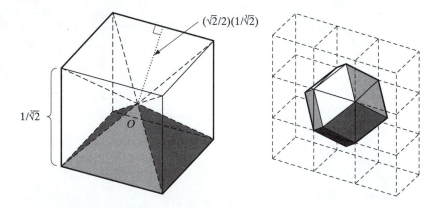

The area of this triangle is $(1/2)(1/\sqrt[3]{2})(\sqrt{2/2})(1/\sqrt[3]{2}) = 2^{5/6}/8$. Thus, the surface area of the rhombic dodecahedron is $(3)(2^{5/6}) \sim 5.345$.

Truncated Octahedron. Finally, let's calculate the surface area of a truncated octahedron of edge length s. The surface consists of six squares of side s (total area $6s^2$) and 8 regular hexagons of side s (total area $12\sqrt{3}\,s^2$). Thus, the surface area is $(6 + 12\sqrt{3})s^2$.

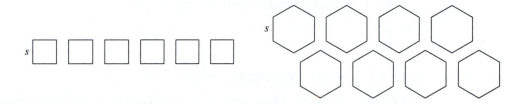

The truncated octahedron of volume 1 in the picture below is half the volume of the cube it sits in. Thus this cube has volume 2 and edge length $\sqrt[3]{2}$. Consequently the edge length of the truncated octahedron of volume 1 is $s = (\sqrt{2}/4)(\sqrt[3]{2})$.

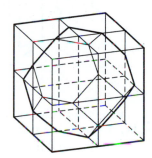

Putting all this together, we find that the surface area of the truncated octahedron of volume 1 is $(6 + 12\sqrt{3})(1/8)(\sqrt[3]{2})^2 = (3/4)(1 + 2\sqrt{3})(\sqrt[3]{2})^2 \sim 5.327$.

To sum up:

- The cube of volume 1 has surface area 6
- The rhombic dodecahedron of volume 1 has surface area about 5.345
- The truncated octahedron of volume 1 has surface area about 5.327

Both the rhombic dodecahedron and the truncated octahedron have less surface area than the cube of the same volume. Of these two, however, the truncated octahedron wins, but not by much, even though the density of the sphere packing associated with it is less (.68) than that of the sphere packing associated with the rhombic dodecahedron (.74). This is not at all what we expected. It suggests that there might be an alternative cap to the end of a bee cell that fits all the requirements but that uses less wax than the rhombic dodecahedron cap. It turns out that there is such an alternative. See Project M5 in Chapter 9 for more details on this.

The geometries of plane and space appear to be quite different. A final solution to the isoperimetric problem for space filling polyhedra has not been found. Add to this the fact that the problem of finding the densest sphere packing has not been solved either. Our understanding of space geometry seems to be much sketchier than our understanding of plane geometry.

✳ *The Cubic Close Packing Space Grid*

Consider spheres in simple cubic packing. Imagine connecting the centers of neighboring spheres with toothpicks and having the spheres themselves disappear. You will wind up with a "space" grid looking like the skeletons of space-filling cubes.

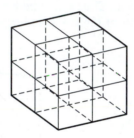

Now think of spheres in cubic close packing. Imagine connecting the centers of neighboring spheres with toothpicks. Then throw away the spheres. What you will have left is a space grid. We'll figure out what it looks like in the following experiment.

Experiment 12   Take a bunch of toothpicks and some soft objects with which to form joints. (Chick pea size pieces of Play-Doh™ would work. Or modeling clay. Or raisins!) Start with the first layer of spheres in the triangular arrangement in a plane. The spheres have centers at the points O talked about in the beginning of this discussion. Assume the spheres of the next layer have centers above the points B and the spheres in the third layer have centers above the points R.

Place raisins at the O points in the plane and connect neighboring O points with toothpicks.

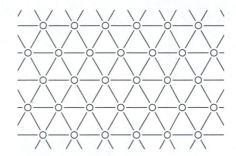

Next connect the B points in layer 2 with neighboring O points in layer 1. This will create the skeletons of a bunch of regular tetrahedra each with three O vertices and one B vertex.

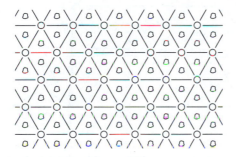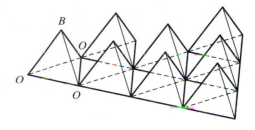

Finally, connect neighboring points B with themselves.

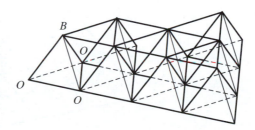

The connections between three neighboring Os and the three neighboring Bs directly on top form the following skeleton.

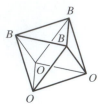

It's a regular octahedron! A regular octahedron fits exactly in the space between three tetrahedra arranged in a triangle. The space grid formed by connecting two layers of cubic close packing consists of alternating skeletons of regular tetrahedra and octahedra.

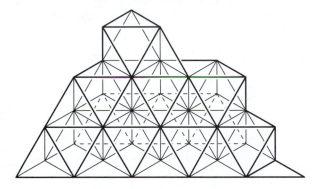

Connecting the B points of the second layer with neighboring R points of the third will create regular tetrahedra sitting on top of each octahedra and oriented in the same way as the tetrahedra of the first layer, but offset the same amount as a sphere of the second layer is offset from a neighboring sphere of the first.

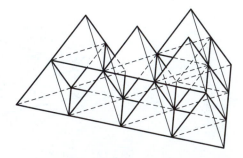

Then, connecting the neighboring R points of the third layer will create a new bunch of regular octahedra, each one directly above an O of the first layer.

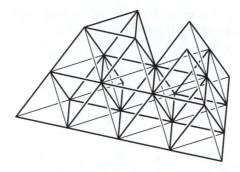

Finally, adding a fourth layer and making all the connections, there will be regular tetrahedra sitting directly above the O points and regular octahedra sitting directly above the B points. After this the pattern repeats. We wind up with a skeleton of regular tetrahedra and octahedra that alternate to fill space. Extraordinary!

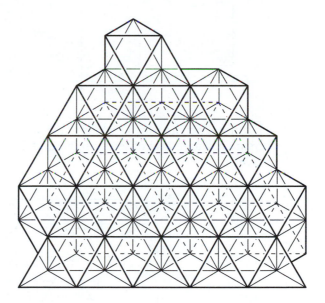

✳ The Cubic Close Packing Space Grid from the Cannonball Perspective

Another way to see the octahedra and tetrahedra associated with cubic close packing is to look at the cannonball arrangement. The spheres on the first layer are in the square arrangement of circles in the plane. Call the centers of these spheres P. Call the centers of the squares Q. (The centers of spheres in the second layer lie above the Q points of the first layer.) We'll construct the space grid in the following experiment.

Experiment 13   With raisins at P points in the first layer, use toothpicks to connect centers of neighboring P points.

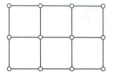

Next, connect each Q point in the second layer to its neighboring P points on the first layer. What you will get is an array of Egyptian pyramids, each being half of a regular octahedron.

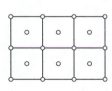
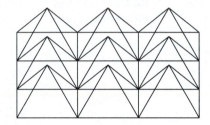

Then, connect neighboring Q points in the second layer. In the second layer this will create a square grid offset from the grid determined by the P points on the first layer.

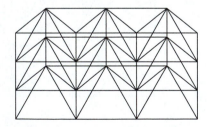

More surprising are the shapes formed between the half octahedra. Take a pair of neighboring Q points in the second layer. From our bird's eye view the connection joining these points appears to be perpendicular to the connection joining two P points on the first layer just below it.

The connections between the Q points and the P points on the two "perpendicular" lines create a regular tetrahedron! Another surprise! (I hope you haven't overdosed on surprises, yet.)

(You may recognize this view of a regular tetrahedron from Chapter 6.) Finally, the space between four of these tetrahedra—determined by the four vertices of a grid square on the second layer—is an inverted Egyptian pyramid, its apex at a P point in the first layer. Of course, this half octahedron will be "completed" when connections are made from the four Q vertices of the second layer to the neighboring P point of the third layer.

Problem 28

You recall that there are two ways a regular tetrahedron sits snugly inside a cube. Call these two ways R and B. Fill space with cubes and color them red and blue, checkerboard style. In each red cube inscribe an R tetrahedron and in each blue cube inscribe a B tetrahedron. In this way we get space filled by tetrahedra and the stuff that fits between them. What is this stuff? Is there any connection between the arrangement of tetrahedra here and the arrangement of tetrahedra in cubic close packing?

Problem 29

Construct the space grid corresponding to body centered cubic packing. You might want to try this using The Geometry Files. (The text suggests a couple of ways of doing it.)

Problem 30

Construct the space grid corresponding to hexagonal close packing. You might want to use The Geometry Files.

Problem 31   Through each sphere of the cubic close packing of spheres there are seven distinct planes of mirror symmetry. Four of these are planes of symmetries of spheres in the square arrangement in the plane. Three are planes of symmetries of spheres in the hexagonal arrangement in the plane. Explain how they come about and describe how they are related to each other.

Problem 32   Earlier we counted the number of wires surrounding a central wire. (See problem 11.) By analogy, surround the close-packed cuboctahedron with more layers. How many spheres altogether?

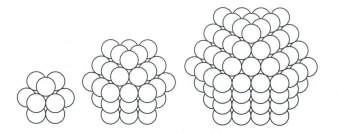

Problem 33   Fill space with cubes in the usual way. Place spheres at each vertex and in the middle of each cube face so that the latter touches all four spheres located at the vertices of that face. This is called **face centered cubic packing.** Show that this is the same as cubic close packing.

✳ *Full Circle*

In this chapter, we considered two main questions, one analogous to the other. First, in the plane: How can we arrange circles so that the amount of wasted space is minimal? Second, in space: How can we arrange spheres so that the amount of wasted space is minimal? These questions of efficiency continue an earlier theme of the book. These questions led us to look at another pair of analogous questions. In the plane: Of all polygons which tessellate the plane, which has the least ratio of perimeter to area? In space: Of all polyhedra which tessellate space, which has the least ratio of surface area to volume?

Both problems in the plane have solutions. We solved the problem with circles using Jensen's Inequality. The solution to the isoperimetric tessellation problem can be found in the reference by Niven. Since the resolutions to both problems involve the tessellation of the plane by regular hexagons, we wondered if they accounted for the prevalence of hexagonal arrays in nature—such as in bee cells. One outcome of this side investigation was a theorem—a consequence of Euler's formula—giving six as the average number of sides of polygons filling the plane.

Neither problem in space has a solution—yet. However, we gathered a lot of evidence and we know that what we might expect by analogy with two dimensions is not true. We have insight, partial success, examples, but no prize. This is the nature of mathematics: There are always intriguing, unsolved problems around. And when those problems are solved, there will be more questions and more problems to solve.

✳ *Notes*

The structure of a honeycomb in a beehive has been an object of fascination for a long time. Aristotle admired and discussed it. Concerning the possible geometric configurations the bees might have chosen, Pappus of Alexandria (ca. 300 A.D.) wrote:

> Now only three rectilineal figures would satisfy the condition, I mean regular figures which are equilateral and equiangular; for the bees would have none of the figures which are not uniform . . . . There being then three figures capable by themselves of exactly filling up the space about the same point, the bees by reason of their instinctive wisdom chose for the construction of the honeycomb the figure which has the most angles, because they conceived that it would contain more honey than either of the two others. Bees, then, know just this fact which is of service to themselves, that the hexagon is greater than the square and the triangle and will hold more honey for the same expenditure of material used in constructing the different figures. [Heath (1921), p. 390]

Kepler may have been one of the first to explain the hexagonal cross-section and end face structure of the honeycomb tubes on the basis of packing tubes subject to internal pressures. D'Arcy Wentworth Thompson (1961) supported this argument, saying that it makes more sense to suppose "that the beautiful regularity of the bee's architecture is due to some automatic play of the physical forces" than to suppose "that the bee intentionally seeks a method of economizing wax." Niven (1981) (see Chapter 8) includes a proof that of all polygons of fixed area that tessellate the plane, the regular hexagon is the one that has the least perimeter. Fejes Tóth (1964), in his article "What the Bees Know and What They Do Not Know" has shown that the cap the bees have chosen for the ends of their cells does not use the least amount of wax.

Hilbert and Cohn-Vossen (1952) provide an alternative argument to the one we give that the hexagonal arrangement of circles in the plane is the arrangement of greatest density. Their argument assumes that the centers of circles are *lattice* points for some lattice in the plane. (A lattice in the plane is the set of all points of the form $a\mathbf{v} + b\mathbf{w}$ where $\mathbf{v}$ and $\mathbf{w}$ are fixed nonzero vectors in different directions and a and b are integers. The square and triangular lattices of Chapter 1 are examples.)

Dirichlet regions (also known as Voronoi regions) play a role in a simple model of crystal growth which starts with a fixed collection of sites in 2- or 3-space, and allows crystals to begin growing from each site, spreading out at a uniform rate in all directions, until all space is filled. The "crystals" then consist of all points nearest to a particular site, and consequently are just the Dirichlet regions for the original set of points.

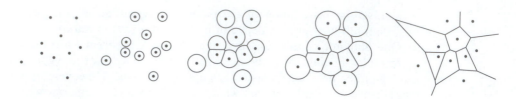

Model of two-dimensional crystal growth

In 1611 Johannes Kepler conjectured that the cubic close packing of spheres is the densest packing of spheres in space. A related problem—How many spheres can be arranged around a central sphere in such a way that all the surrounding spheres just "kiss," or touch, the central sphere?—was the subject in 1694 of a famous dispute between Isaac Newton and David Gregory, a Scottish astronomer. This problem wasn't settled until 1953 by K. Schütte and B. L. van der Waerden in Newton's favor. The answer is 12.

One experiment to find the best packing of spheres was reported in 1727 by Stephen Hales in his *Vegetable Staticks,*

> I compressed several fresh parcels of Pease in the same Pot, with a force equal to 1600, 800, and 400 pounds; in which Experiments, tho' the Pease dilated, yet they did not raise the lever, because what they increased in bulk was, by the great incumbent weight, pressed into the interstices of the Pease, which they adequately filled up, being thereby formed into pretty regular Dodecahedrons. [Coxeter (1969), p. 409]

In 1831, Gauss showed that cubic close packing is densest assuming that the centers of the spheres constitute the points of some lattice in space. An argument for this is also contained on p. 45 of Hilbert and Cohn-Vossen's book.

Until recently, no general proof was found that cubic close packing is densest. However, in 1990 Wu-Yi Hsiang of the University of California at Berkeley announced a proof of Kepler's conjecture. At the time this book went to press, the proof was being challenged by the mathematical community. See the articles by Hale (1994) and Hsiang (1995) for more details of this controversy.

Buckminster Fuller felt that the cubic close packing of spheres represented the shape of space. He called the associated space grid the **octet truss.** Pieces of the space grid are used as structural elements in building and construction. Some versions are shown below.

Left: *Plaza Côtes des Neiges.* Shopping center, Montreal. Architects: Mayers and Girvan. Structural engineers: Baracs and Gunther. Right: "Octet" spaceframe (student project).

Sphere packings have been studied for many years in part because of their broad implications for understanding the behavior of solids and liquids. For example, knowledge of sphere packing can provide new insights into the ways atoms can be packed in materials such as superconductors. Equally important is the application of how spheres pack to the properties of powders and porous materials.

Molecular properties of many crystalline materials can be described using sphere packing in some way. The atoms of many metals such as copper arrange themselves as if they were closely packed spheres. Sometimes the atoms are arranged in hexagonal close packing (ORORORO . . .), sometimes cubic close packing (ORBORBORB . . .).

Body-centered cubic packing of atoms in
sodium, lithium, potassium and molybdenum crystals.

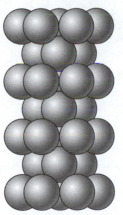
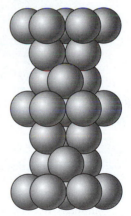

Hexagonal close packing of atoms in
magnesium crystals.

Cubic close packing of atoms in
copper, silver and gold crystals.

When a molecule is made up of atoms of more than one element, in many of them the atoms arrange themselves so that the atoms of one element are close packed while the others occupy the curvy tetrahedral or octahedral interstices. Here is an example.

In sodium chloride (NaCl), common salt, the chloride (Cl) ions are arranged in cubic close packing; the sodium ions (Na) fill up octahedral interstices. Conveniently, there is one octahedral cavity for each sphere in the packing. (This can be seen in the cannonball perspective on the cubic close packing space grid.)

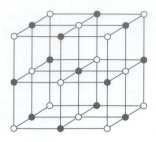

The following pictures—electron micrographs—of even more complex molecules suggest that the molecules themselves are arranged in some kind of close packing.

 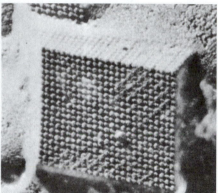

Electron micrographs of crystals, showing arrays of individual molecules. Protein from southern bean mosaic virus (magnification 30,000) (*left*). Protein from tobacco necrosis virus (magnification 73,000) (*right*). Photographs by R. W. G. Wyckoff.

To end this discussion on the uses of arrangements of spheres in space, here is a tantalizing quote from N. J. A. Sloane (1984) about issues in accurate data transmission: "Quantizing data from a continuously variable source is closely related to the problem of covering space [in many dimensions!] with the least dense arrangement of overlapping spheres."

✳ *References*

Coxeter, H. S. M., *Introduction to geometry.* New York: Wiley, 1969.

Edmondson, Amy C., *A Fuller explanation: The synergetic geometry of R. Buckminster Fuller.* Boston: Birkhauser, 1987.

Fejes Tóth, L., *Regular figures.* New York: Macmillan, 1964.

Fejes Tóth, L., "What the bees know and what they do not know," *Bulletin of the American Mathematical Society,* vol. 70 (1964), 468–481.

Hale, Thomas C., "The status of the Kepler conjecture," *Mathematical Intelligenser,* vol. 16, no. 3 (1994), 47–58.

Heath, Thomas, *A history of Greek mathematics,* vol. 2, New York: Oxford University Press, 1921.

Hilbert, David, and S. Cohn-Vossen, *Geometry and the imagination.* New York: Chelsea Publishing Co., 1952.

Holden, Alan, *Shapes, space and symmetry.* New York: Columbia University Press, 1971.

Hildebrandt, S., and A. Tromba, *Mathematics and optimal form.* New York: Scientific American Library. Distributed by W. H. Freeman, 1985.

Hsiang, Wu-Yi, "On the sphere packing problem and the proof of Kepler's conjecture," *International Journal of Mathematics,* vol. 4, no. 5 (1993), 739–831.

Hsiang, Wu-Yi, "A rejoinder to Hale's article," *Mathematical Intelligencer,* vol. 17, no. 1 (1995), 35–42.

Kappraff, Jay, *Connections: The geometric bridge between art and science.* New York: McGraw-Hill, 1991.

Lines, L., *Solid geometry.* New York: Dover, 1965.

Loeb, Arthur L., *Space structures—their harmony and counterpoint.* Reading, MA: Addison Wesley, 1976.

Niven, Ivan, *Maxima and minima without calculus.* Washington D.C.: Mathematical Association of America, 1981.

O'Daffer, Phares, and Stanley R. Clemens, *Geometry: An investigative approach.* Menlo Park, CA: Addison-Wesley, 1997.

Senechal, Marjorie, and George M. Fleck, *Shaping space: A polyhedral approach.* Boston: Birkhauser, 1988.

Shubnikov, A. V., and V. A. Koptsik, *Symmetry in science and art.* New York: Plenum Press, 1974.

Sloane, N. J. A., "The packing of spheres," *Scientific American,* January 1984, 116–125.

Steinhaus, Hugo, *Mathematical snapshots.* New York: Oxford University Press, 1950.

Stevens, Peter S., *Patterns in nature.* Boston: Little, Brown, 1974.

Thompson, D'Arcy Wentworth, *On growth and form.* Cambridge: Cambridge University Press, 1961.

9 Where to Go from Here? Projects

Now that you've been through this book, or most of it, what's next? One thing is to step out on your own, find a problem or a topic related to things we've talked about here that interests you, find a group of chums with the same interests (if you want company), go to town, and have some fun. One purpose of this chapter is to give you some suggestions. So the bulk of the chapter is just that: a list of project ideas. After you wrestle with the project for awhile and you feel like you're an expert on a topic or you've solved some problems or you have some great insights, it's great to come back to the rest of the world and share what you found. A second purpose of the chapter is really about this aspect of project-doing: communicating your findings. There's nothing worse than telling someone what you've done and having them fall asleep or roll their eyeballs. You'd like to communicate the results of your project effectively.

✳ Project Ideas

Here is an annotated list of ideas for projects. The ideas span a wide range of project types. At one end, there are exploratory projects—similar to the mini-projects found in Chapter 1; these are basically problems to solve; no references are given for these. At the other end of the spectrum, are expository projects. Such a project involves taking a known geometric procedure, idea, technique, fact (or collection of related facts), or theorem. Then the material would be put together in a new way, relating it to things that happened in the book and to previous experiences with geometry and making a model (or two) that illustrates the material. There are lots of references to the expository projects. Most projects will be some combination of the two. On the one hand, solving a problem sometimes leads to finding out what others have done. On the other hand, trying to understand what others have done in an area you're interested in sometimes unexpectedly leads to solving new problems.

In the list below, the annotations are not uniform. For some project ideas, there are lots of details. For others, there's just a title, a few questions, and a reference or two. This uneven presentation may not only give you ideas but also show you ideas in various stages of development—some with just the germ, some with lists of questions to ask (and answer) and suggestions for things to make, others with a detailed outline.

Sometimes the references are sketchy. They are suggestions; they are not complete. I hope that what I list here will get you started and that you will find what you need on your own.

There are lots of good project ideas not on this list! Items in the text may suggest ideas directly to you. If you don't find something below that excites you, use the list to get yourself thinking. An idea might suggest something else to you. The references at the end of the list include many books for browsing. Talk with friends and teachers. Happy hunting!

A. Tessellations of the Plane

A1. Tiling with regular polygons. In Chapter 1, the tiling problems with regular polygons all assume that every vertex is the same. What if you relax this requirement? [Senechal (1988b); Grünbaum and Shephard (1987); Kappraff (1991), p. 183ff]

A2. Tiling with nonregular polygons. In Chapter 1, there are problems involving tiling with triangles and quadrilaterals. Try using nonregular convex pentagons, hexagons, ... What can you say? What about nonconvex polygons? [Niven (1981), pp. 149ff; Senechal (1988b); Grünbaum and Shephard (1987)]

 A3. Creating new tiling patterns out of old. Perhaps the most famous tiling patterns are those created by the Dutch graphic artist M. C. Escher. You can create some of these yourself out of simple patterns you know. [Kappraff (1991), pp. 234ff; Ranucci and Teeters (1977)] In addition, you might want to check out *TesselMania*®, computer software that utilizes ideas from these sources.

A4. Nonperiodic tiling. In Chapter 1 there are problems involving tiling by polygons that produce patterns that repeat to produce a wallpaper design. Some tiles that produce repeating designs also produce nonrepeating (i.e., nonperiodic), yet plane-filling designs. Some do this in many different ways. [Kappraff (1991), pp. 194ff; Grünbaum and Shephard (1987); Senechal (1988b); Fejes Tóth (1964)]

A5. Penrose tiles. This is another kind of non-periodic tiling involving two kinds of tiles: darts and kites. The patterns formed have remarkable properties. [Steen (1988), Chap. 13; Senechal (1988b), pp. 3–22+; Senechal (1996); Grünbaum and Shephard (1987); Kappraff (1991), pp. 194ff, 243ff; Nelson (1985)]

A6. Copies of a certain shape may make a tiling, but some copies may be rotations or flips of the original shape. Which shapes will tile by using only translations of the original shape? [Grünbaum and Shephard (1987); Senechal (1988b)]

B. Two- and Three-Dimensional Lattices

B1. Shooting bullets through the plane. Imagine the upper-right quadrant of the plane as an infinite piece of graph paper, whose squares are painted black and white like a checker board. Imagine also shooting a bullet from the origin in a straight line through the quadrant. What you want to know is this: What percentage of the time in the trip does the bullet spend passing through the black squares?

B2. Shooting bullets through space. Imagine the positive octant of space tiled with cubes painted black and white, like a three-dimensional checkerboard. Imagine shooting a bullet from the origin in a straight line through the octant. What percentage of time in the trip does the bullet spend passing through black cubes? Compare with project B1.

B3. Billiard tables, revisited—compare with Mini-Project 12 of Chapter 1. (Thanks to Fred Stevenson.)

(a) Take an $n \times m$ billiard table where n and m are positive integers. Shoot a billiard ball out of the corner at an angle Θ. What can you say?

(b) Take a $1 \times \sqrt{2}$ billiard table. Shoot a ball out the corner at an angle of $45°$. What can you say? (How big should pockets be at the corners so that the ball enters a pocket in 1,000,000 rebounds off the sides, but not before?)

B4. Three-dimensional billiards. You have an $n \times m \times p$ rectangular box with mirrors on all its faces. From one corner, and in the direction of a cube's diagonal, you shoot a laser beam. What can you say about the path of the laser beam? Compare with Mini-Project 12 of Chapter 1.

B5. Squares on the three-dimensional integer lattice. Using integer lattice points in the plane as vertices, you can create a variety squares. See Mini-Project 4 in Chapter 1. What sorts of squares can you place in space such that their vertices are three-dimensional integer lattice points? What are their areas? Are there squares in space whose areas can't be obtained using integer lattice points in the plane?

B6. Cubes on the three-dimensional integer lattice. What sorts of cubes can you place in space such that their vertices are three-dimensional integer lattice points? What are their volumes? Compare with project B5. (This is Mini-Project 5 of Chapter 1 in disguise. If you didn't try it there, you might want to try it here!)

B7. Dirichlet domains of lattices and their relation to plant growth. Pineapples. Pine cones. Sunflowers. [Peressini (1980); Coxeter (1969), p. 189; Kappraff (1991), p. 238ff]

C. Measurement

C1. How the Greeks found various areas and volumes. How did Euclid deal with the area of a circle? How did Archimedes deal with the surface area of a sphere? How did Euclid deal with the volume of a pyramid? [Heath (1953, 1956); Katz (1993)]

C2. Measuring the earth and the solar system. How did the ancients measure the circumference of the earth? The distances of the earth to the moon and sun? The diameters of the moon and sun? [Rickey (1992); Layzer (1984); Steen (1988), Chaps. 16, 17; van Heldon (1985)]

C3. The orbital geometry of the earth, sun, and moon. How can you use geometry to predict eclipses of the moon and sun?

C4. The length of the day. Can you predict it? How does it vary throughout the year? What does it have to do with the geometry of the earth, the earth's orbit about the sun, the earth's tilt, and the earth's rotation? [Bassein (1992); Wagon (1990)]

C5. Sundials. How to make one? How do they work? Make one. [Waugh (1973)]

C6. On being the right size. Could an animal like King Kong really exist? Is there a limit to the size of a mountain? Are there limits to the size of a building? This is a sketchy project idea; it needs more focus. Deals with how area and volume of an object (or organism) change as linear scale changes. Deals also with connections between these area and volume measurements and other physical properties, constraints, and requirements of the object. [Steen (1988); Haldane "On Being the Right Size," in Newman (1956); McMahon and Bonner (1983); D'Arcy Wentworth Thompson (1961, Chapters I and III.)]

D. Polyhedra

D1. The semiregular polyhedra, I. Find them all. Show why your list is complete. Make models of them. First hint: There are two infinite families of semiregular polyhedra plus a finite set of "exceptional" semiregular solids (sometimes called Archimedean solids). Second hint: Regular polyhedra correspond to tessellations of the sphere by a single regular spherical polygon; the latter are analogous to tilings of the plane by a single regular polygon. Which tilings of the sphere correspond to the semiregular polyhedra and what are the analogous tilings in the plane? [Lines (1965); O'Daffer and Clemens (1997), Chap. 4; Pearce (1978); Fejes Tóth (1964)]

D2. Semiregular polyhedra, II. Take the list of the exceptional semiregular solids in D1 and show how they are related by truncation (or other types of polyhedral alteration) to themselves and to the regular polyhedra. (This is a way of showing that the semiregular solids you think exist really do.) You might also want to check out appropriate files from The Geometry Files (or create your own) in order to understand or illustrate what is going on. What are the analogous alterations to (and relationships among) tessellations in the plane? [Lines (1965); O'Daffer and Clemens (1997); Senechal (1988b); Pearce (1978); Fejes Tóth (1964)]

D3. Semiregular polyhedra, III. Of particular interest are two unusual semiregular polyhedra, the Snub Cube and the Snub Dodecahedron. Do they exist? How do you construct them? [Lines (1965); O'Daffer and Clemens (1997); Senechal (1988b); Pearce (1978); Fejes Tóth (1964)]

D4. "Efficient" polyhedra. See project L2.

D5. Classify polyhedra of a certain type. Examples: Find all polyhedra with triangular faces (subproblem: Find all **convex** polyhedra with equilateral triangular faces); find all the combinatorial types of polyhedra with 12 or fewer faces; find all polyhedra with pentagonal faces; find all polyhedra with pentagons and hexagons as faces (see also project D8); find all polyhedra whose faces are quadrilaterals [Beck, Bleicher, and Crowe (1970); Senechal (1988b); Loeb (1976), p. 38].

D6. Equidecomposability of polyhedra (Hilbert's third problem). In the plane, if you take two polygons having the same area you can cut one of them up into a finite number of smaller polygonal pieces and reassemble the pieces to form the other. What happens in space? Of two polyhedra in space with the same volume it is not always possible to cut one up and reassemble the pieces to form the other. The proof of the first statement is accessible to any high school student. The proofs (that I know of) of the second statement need some abstract, linear algebra-type reasoning. It would be nice to find (in the literature) or to create a more accessible proof. Project should show, in an understandable way, that the methods we all know of finding areas of polygons in the plane—by dissection, rearrangement, embedding, relating to area of rectangle—will not work for finding volumes of polyhedra. Other, strikingly different, methods are needed. Project should also mention the history of the problem. [Kaplansky (1977); Boltyanskii (1963); McCrory (1986)]

D7. Polyhedra related to the regular icosahedron. There are several polyhedra whose faces are equilateral triangles (or almost so) all of whose vertices are of order 5 or 6. What are they and how are they related to the regular icosahedron? You might want to construct some of these using The Geometry Files. Buckminster Fuller has proposed that these shapes be used as building structures. What makes these shapes so desirable? What is an *n*-frequency icosahedron? You might want to make models of some of these. [Senechal (1988a); Loeb (1976), pp. 36; Edmondson (1987), pp.78; Hildebrandt and Tromba (1985)]

D8. Polyhedra related to the regular dodecahedron. Consider polyhedra all of whose faces are pentagons and hexagons, all of whose vertices are of order 3. These are related to the regular dodecahedron. How? They are also related to the shapes described in project D7. How? Some of these shapes occur in nature. Where? Why? What properties of these dodecahedral-like polyhedra make them a desirable shape in nature? In Hildebrandt & Tromba (1985, p. 176) the authors suggest that the icosahedral-like form of radiolaria (microscopic sea animals) is related to the 90° and 120° rules for soap bubbles. See also project D7. [Senechal (1988a); Loeb (1976), p. 36; Edmondson (1987), pp. 78, 117; Hildebrandt & Tromba (1985); Kappraff (1991), p. 323]

D9. Buckyfullerine. A new molecule of carbon has been discovered/created in the last decade. It's called Carbon 60 or Buckyfullerine or just plain Buckyball. It's related to both the soccer ball and to geodesic domes. What is its geometric structure? What properties does it have and how are they related to its geometric structure? How is it that it got the geometric structure that it has? There are other, large, related molecules of carbon. What are they? [Saunders (1991); Schmalz (1988)]

D10. Tessellations of space with polyhedra. See projects E1–E5.

 D11. Four-dimensional "polyhedra." What are the four-dimensional analogues to regular polygons in the plane and regular polyhedra in space? What should a definition of such an object be? Construct three-dimensional projections of these. Or create projections of these using three-dimensional computer graphics software such as The Geometry Files. [Banchoff (1990); Banchoff (1978); Loeb (1976), p. 58+; Senechal (1988a); Fejes Tóth (1964)]

D12. Tetrahedral "rings." You can make several closed rings out of tetrahedra. You can turn some of them "inside out." What are they? Classify them. Show that your list is complete. Make models of them. One ring will fold up into a cube. [Ball and Coxeter (1962); Schattschneider (1990); Kappraff (1991), pp. 279ff]

D13. Star polyhedra. You can relax the definition of a polygon so that it includes the "star polygon," shown below.

What should the definition of a star polygon be? The definition of a regular polyhedron might also be relaxed to allow its faces to be regular star polygons. What should the definition a "star polyhedra" be? What are some examples? Kepler included a discussion of the "stellations" of the regular polyhedra in his treatise on polyhedra. What is the definition of a stellation of a polyhedra? What are the stellations of the regular polyhedra? Make models. [Senechal (1988b); Senechal (1988a); Cundy and Rollett (1961); Cromwell (1995); Kappraff (1991), pp. 288ff]

D14. Origami polyhedra. Make origami models of several of the regular and semi-regular polyhedra. Design your own. Create methods for some that haven't yet been made (that you know of). [Kappraff (1991), pp. 198+; Fusè (1990); Kasahara and Takahama (1987)]

 D15. Nested regular polyhedra. Make the following set of nested regular polyhedra: Octahedron fits inside tetrahedron so that every vertex of former is the mid-points of an edge of latter; tetrahedron fits inside cube so that every edge of former is diagonal of face of latter; cube fits inside dodecahedron so that every edge of former is diagonal of face of latter; dodecahedron fits inside of icosahedron so that every vertex of former is center of face of latter. To build this you will need to find ratios of edge lengths. You might also want to construct this using The Geometry Files. Find other interesting nestings. Compare with Kepler's nesting of the regular polyhedra. [Hopley (1994); Beard (1973)]

E. Tessellations of Space

E1. Space-filling tetrahedra. Some tetrahedra will tile space. (The regular tetrahedron is not one of them.) What are they? Build models. [Senechal (1981)]

E2. Space-filling octahedra. There are octahedra which tile space. (The regular octahedron is not one of them . . .) What are they? Build models. [Senechal (1981); Loeb (1976), pp. 148]

E3. Space-filling shapes. What single polyhedra fill space? Suppose translations of a single shape fill space. What are the possibilities for such a shape? Find arguments. Build models. (Compare with projects E1 and E2.) [Senechal (1981); O'Daffer and Clemens (1997), Chap. 4]

E4. Semiregular tessellations of space, I. Tile space using combinations of polyhedra chosen from among the semi-regular and regular solids. What are the possibilities? Build models. Construct them using The Geometry Files. [O'Daffer and Clemens (1997), Chap. 4; Pearce (1978)]

E5. Semiregular tessellations of space, II. What connections can be made between the semiregular tessellations? (See project E4.) Many tessellations of the plane by regular polygons (single polygon or a combinations of several different polygons, but all vertices look alike) are related by truncation and other alterations. Many regular/semiregular polyhedra are related by truncation and other alterations. These space tessellations may be related in similar ways. What are they? These connections may help to prove existence of some of the tessellations found in project E4. See similar discussions in projects D1 and D2. Use The Geometry Files to show the relationships. [O'Daffer and Clemens (1997), Chap. 4]

F. Rigidity

F1. Take a rectangular grid in the plane. What kind of bracing will make it rigid? Make some frameworks and experiment with them a bit. Afterwards, look at Baglivo and Graver. Next, suppose that you construct a three-dimensional grid, a sort of stacking of cube skeletons. What kind of bracing would be necessary to make it rigid? [Baglivo and Graver (1983); Grünbaum and Shephard (1988); Chakravarty (1986); Edmondson (1987), pp. 55ff; Loeb (1976), pp. 29–30; Kappraff (1991), pp. 154ff]

F2. Stability of polyhedral skeletons, I. Make a polyhedral skeleton. What does it mean for the skeleton to be "stable"? Loeb states this general rule: If a three-dimensional polyhedral skeleton is stable, then $E = 3V - 6$. Prove this. Is the converse true? Are there exceptions? [Loeb (1976), pp. 30; Kappraff (1991), pp. 270ff]

F3. Stability of polyhedral skeletons, II. Rule: A polyhedral skeleton consisting of triangular faces must be stable; i.e., it doesn't need interior bracing. What about Fuller's "dimples"? What about the flexahedra? What's going on? [Loeb (1976), p. 5; Edmondson (1987), pp. 5, 63; Connelly (1979)]

G. Shortest Path Problems

G1. Variations on the Three Cities Problem. In Steinhaus, an alternative, dynamical solution is given. Build the model that goes with it! Also, he solves a similar problem in which the population of the city is accounted for. Describe the solution. Could the latter be generalized to more than three cities? [Tong (1995); Steinhaus (1950), pp. 116ff]

G2. *n*-Cities Problem. Generalize the Three-Cities Problem. How many new "nodes" must/can be added? What can a vertex "look like"? Multiple solutions. See project I4. [Courant and Robbins (1973); Steinhaus (1950); Niven (1981); Hildebrandt and Tromba (1985); Bern and Graham (1989)]

H. Conic Sections

H1. Mirrors, lenses and other uses for ellipses, parabolas, and hyperbolas. It turns out that the hyperbola has a reflection property. What is it? Why is it? How can it be used? What are some uses for the conic sections and their reflection properties? [Whitt (1982); Maesumi (1992)]

H2. Descartes and the Greeks. The Greek and the analytic geometry definitions of a conic section are quite different. Reconcile the two. Dandelin spheres. [Jennings (1994); Courant and Robbins (1973)]

H3. Greeks knew that ellipses and parabolas had reflection properies. What were their arguments? [Katz (1993)]

H4. String figures and paper folding. You can make conic sections out of straight lines! This is the basis of string art and some paper folding exercises. How do these techniques work? Why do they produce conics? Make some models! [Row (1966); Broman (1994); Hilbert and Cohn-Vossen (1952); Courant and Robbins (1973); Seymour (1992)]

I. Soap Bubbles

I1. Soap bubble geometry, I. Here are some experiments from which you can look for patterns in the geometry of soap bubbles. Eventually you will want to come up with explanations for these patterns.

You will need some wire and soap bubble solution (recipe: 10 cups water, 1 cup Joy, 3–4 tablespoons glycerine).

(a) You can get a soap film solution to "span" the space formed by a wire frame. For example, make a wire circle a couple of inches in diameter and attach a "wand" to it:

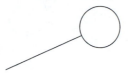

Holding onto the wand, dunk the circle into the solution. What happens?

(b) Another wire form-with-wand to make and dunk in soap solution:

What is the shape of the soap film that forms? Try dunking it several times. What happens?

(c) Physicists claim that the soap film that forms, spanning the space formed by a wire frame, has the smallest possible area of all surfaces that could span the frame. (Such a surface is called a **minimal surface.**) Here is an experiment to try in order to lend plausibility to this claim. Take the wire circle-with-wand that you made earlier. Tie a piece of cotton string (a little larger than the circle's diameter) to the circle's circumference like this:

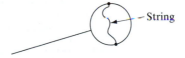

Holding the wand, dunk this in the soap solution. Then with a sharp object, gently "pop" the film on one side of the cotton thread. A minimal surface should form between the cotton thread and the wire frame. (Can you predict, using the physicists' claim, the shape the spanning surface will take?) What shape does the cotton thread take? Can you explain why it is this shape?

(d) Here is another experiment to try. Take your wire circle-with-wand and tie on pieces of cotton thread as in the diagram below.

Holding onto the wand, dunk the wire frame in the soap solution. Gently poke out the "hole"—the film spanning the little cotton loop. What shape does the cotton loop form? Can you explain why?

(e) Some other wire-forms-with-wand to make and dunk:

Make the forms so that the wire is as close together as you can make them—but not touching—at places marked with arrows. When you dunk each one, gently poke out the film spanning the "central" holes using a sharp object. (You may have to try this several times.) Describe what you get.

The following activities don't involve wire frames. All you will need is a flat, glass-like surface (such as a formica table or counter top or piece of Plexiglas™). You will also need a piece of Plexiglas™ (at least a foot on a side) and four bottle corks of the same height.

(f) Wet the glass-like surface. Blow a bubble and gently lay it on the flat surface. Observe where the bubble comes in contact with the flat surface. Blow a few more bubbles and try the same thing. What do you observe? What does the original bubble become? How does the size of the original bubble compare to what you get when you place the bubble on the flat surface? (What is the angle between the bubble and the flat surface?)

(g) Take the rectangular piece of Plexiglas™. Use the four corks to stand it on its corners on the wetted, glass-like surface.

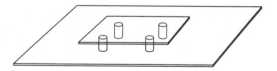

Now blow several bubbles between the Plexiglas™ and the glass-like surface. Look at the bubbles through the Plexiglas™. You should see a kind of honeycomb of shapes. How many "sides" do individual bubble shapes have? How many of these sides meet at a "point"? What angles do adjacent sides make? Anything else you observe? Try the experiment again. Observations?

(h) Next, put the Plexiglas™ "lid" aside. Blow one bubble and lay it on the flat glass-like surface. Blow a second bubble; gently place it next to the first bubble so that the two touch. How does this "cluster" of two bubbles differ from the two bubbles separately? (You no longer have two spheres. What do you have? What happens where the two bubbles come together? What are the angles between the two bubbles where they come together?) Can you predict what will happen if you perform this experiment again? See if your predictions come true.

(i) Now make a cluster of three bubbles. You no longer have three spherical bubbles. What do you have? Describe what happens where two bubbles come together? Describe what happens where three bubbles come together? Angles? Can you predict what will happen if you perform this experiment again? Try this again to see if your predictions come true.

(j) Make a cluster of four bubbles: make the cluster of three as before, then place the fourth bubble on top of the three. Again, observe the interfaces between bubbles (where bubbles come together). How many soap film "walls" meet at a single line? (A wall is a nearly flat section of soap film.) What is the angle between a pair of walls that meet at a line? (How does this compare with what happened in part g?)

How many walls can intersect at a point? (Is there a maximum to this number?) Take a point where four walls meet and look carefully at the angles formed at the "corner." What do you observe? (In part g. did you ever have four walls meeting together?) [Stevens (1974), p. 190ff; Morgan (1994); D'arcy Wentworth Thompson (1961); Courant and Robbins (1973); Hildebrandt and Tromba (1985)]

I2. Soap bubble geometry, II. It would be nice to have some geometric proofs (based on the minimality assumption mentioned in I1 c. above) to explain some of the phenomena observed in the soap bubble experiments. There are several "rules" to explain: the 90° rule, the 120° rule, the arc-cosine$(-1/3)$ rule and the four edge at a vertex rule. The angle arc-cosine$(-1/3)$ pops up all over the place. For example, it is the dihedral angle of a regular tetrahedron. What's the connection with soap bubbles? Hildebrandt and Tromba suggest a proof of the arc-cosine$(-1/3)$ involving a solution of Steiner's problem on the sphere. This might involve Euler's formula and making a nice set of polyhedra models. [Almgren and Taylor (1976); Hildebrandt and Tromba (1985), pp. 120ff; D'Arcy Wentworth Thompson (1961); Stevens (1974); Isenberg (1978)]

I3. Soap bubbles and the icosahedral form of radiolaria. See project D8. [Hildebrandt and Tromba (1985), pp. 176]

I4. Solving shortest path problems using soap bubbles. Analytic solutions to shortest path problems such as the Four-Cities problem and its generalization, the n-Cities problem, are not always available. Soap bubbles can be used to find them. See project G2. [Courant and Robbins (1973); Hildebrandt and Tromba (1985)]

J. Kaleidoscopes and Mirrors

J1. Octahedral kaleidoscope. This is one of the unusual kaleidoscopes made up of three mirrors which meet in a point. It is described in the chapters on kaleidoscopes and symmetry. Make one of these and a variety of shapes to fit in it. A pattern for it is shown here. If you fit the tetrahedron whose pattern is shown here (scale it smaller than the kaleidoscope itself) into the vertex of the kaleidoscope, then the mirror images of the tetrahedron form a cube.

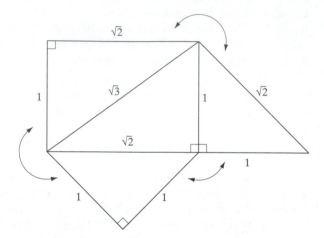

There is another tetrahedron you can fit into the kaleidoscope whose mirror images will make an octahedron. Figure out the pattern and make one. The symmetries of the octahedron and the cube are the same. Find other polyhedra whose symmetries are the same as the cube. For each one construct a shape that will fit into the octahedral kaleidoscope and whose mirror images will make the polyhedron.

J2. Polyhedral kaleidoscopes. The octahedral kaleidoscope of project J1 is intimately connected with the cube and the octahedron. There are other interesting three mirror kaleidoscopes.

(a) Make a kaleidoscope that is to the regular tetrahedron as the octahedral kaleidoscope is to the cube. Construct a shape that will fit "into" the kaleidoscope so that the total design in the mirrors forms a regular tetrahedron. Find other polyhedra whose symmetries are the same as the regular tetrahedron. For each one construct a shape that will fit into the "tetrahedral" kaleidoscope and whose mirror images will make the polyhedron.

(b) Make a kaleidoscope that is to the regular dodecahedron as the octahedral kaleidoscope is to the cube. Construct a shape that will fit "into" the kaleidoscope so that the total design in the mirrors forms a regular dodecahedron. Find other polyhedra whose symmetries are the same as the regular dodecahedron. For each one construct a shape that will fit into the dodecahedral kaleidoscope and whose mirror images will form the polyhedron.

(c) Make other kaleidoscopes out of three mirrors meeting in point (there are infinitely many of these!). Find polyhedra whose symmetries are "generated" by the reflections in the three mirrors and find a shape to fit into the kaleidoscope so that it plus its mirror images form the polyhedron. Hints on these kaleidoscopes are given in the last sections of Chapters 5 and 6. [Ball and Coxeter (1962); Schubnikov (1974)]

J3. Space-filling kaleidoscopes. If you construct a rectangular room with mirrors on all of its walls, floor, and ceiling, then the mirror images of the room's interior fills space. (The total design is all of space! This is analogous to a triangular kaleidoscope where the total design is the whole plane!) There are other mirrored polyhedra whose mirror images fill space. What are these polyhedra? (Some are tetrahedra.) Make some of them. How are they related to other things you know (such as sphere packings of Chapter 8 and the semi-regular space tessellations of E4)? Which "correspond" to cubic close packing, simple cubic packing, body centered cubic packing? [Ball and Coxeter (1962); Schubnikov (1974); Gay (unpublished)]

J4. Cylindrical mirror. Take a paper towel tube and wrap a piece of mirrored mylar around it. This kind of mirror has been around for some time. Like a funhouse mirror, this mirror will distort ordinary shapes. You can draw something warped and misshapen, look at it in this mirror, and its image will appear to be "normal"! How do you produce such drawings? (Such a drawing is called **anamorphic**.) How does such a mirror work? Describe how to make anamorphic drawings. Make some! [Anno (1980)]

J5. Investigate elliptical, parabolic and hyperbolic mirrors: How do they behave, how can they be used? (This is just the germ of an idea. It will need development. You will have to seek out more references.) [Moore (1986); Whitt (1982); Steen (1988)]

J6. MIRAGE. In my office I have a gadget called MIRAGE. It looks like two salad bowls, one inverted and placed on top of the other. There's a hole in the top one about 2 to 3 inches in diameter. A quarter appears suspended in the hole. But when you reach for it, it's not there. How does this work? (Mirrors are involved.) What other properties does this thing have? The toy is manufactured by Opti-Gone Associates Inc., Box 8118, Van Nuys, CA 91409. [Hecht (1987); find more of your own references!]

J7. Experiments with Christmas tree balls. (With references to two-mirror kaleidoscopes discussed in course, including case when the two mirrors are parallel). Take a bunch of spherical Christmas tree balls, all mirrored, all silver in color, all the same size. Arrange them like this, vertically:

(a) Put a hand near this arrangement and look at its reflections in the Christmas tree balls. Are the reflections all different? **How** do they differ? **Why** are they different?

(b) Each Christmas tree ball is a sphere with a mirror **on its outside**. Compare this with the sphere mirrored **on its inside**. What are some differences between a reflected image of a hand on the outside of a sphere and a reflected image of it on the inside? (Explain these differences, if you can.)

(c) In the arrangement above, only a part of each sphere is "illuminated" with reflections. Describe the shape of this part. Look at the unilluminated space **between** spheres, where three come together. Describe the shape of this space. Does this shape surprise you? Why?

(d) Place an illuminated light bulb in the room some distance (10 ft?) from the exhibit and in front of it. The light should be reflected in each of the Christmas tree balls. Describe how the light is reflected in each ball. Does this surprise you? Why? Explanations? (Look closely at the "edge" of the unilluminated space. What do you see? Why?)

(e) If you place an object between the two flat mirrors of a two mirror kaleidoscope, you see reflections of reflections of the object. The same thing happens in a three flat mirrors kaleidoscope. What happens when you place an object between two (or more) spherical mirrors? Does any of this shed any light on explaining the reflections you see in the Christmas tree balls arrangement?

(f) Now arrange the Christmas tree balls this way:

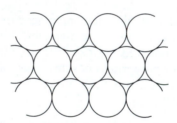

What would be the shape of the illuminated part of each sphere? What would be the shape of the unilluminated space between spheres, where four of them come together? What would be the geometric pattern of light reflections in each ball? Comparisons with the original arrangement? Explanations? [Drucker (1992); Thomas (1980)]

J8. Billiard Balls in Space. See project B4.

J9. Polygonal rooms not illuminable from every point. Actually, this is the title of a paper by Tokarsky (1995). In it he talks about the problem of two people being in a dark room (which might have many turns and cul-de-sacs) where the walls, floors, and ceilings are mirrored. One lights a match. Can she see the other person, no matter where the two are located? [Tokarsky]

J10. Hall of Mirrors: a maze so bewildering that the explorer must rely on a map. A paraphrase of a title of another article, by Walker (1986). Mirrored labyrinths. How do they work? How to get out of them? How to make one? [Walker]

K. Symmetry

K1. Two- and three-dimensional frieze patterns. What are they? Give a classification of them by symmetry. Construct models of the three-dimensional patterns. [Shubnikov describes the two- and three-dimensional patterns; Fejes Tóth (in *Regular. . .*) describes and discusses the two-dimensional frieze patterns]

K2. Wallpaper designs. Extend the investigation of the symmetries of wallpaper designs begun in the text. If you classify designs by symmetry, then there are only 17 possible designs. The text includes one from each class. Why are there just 17? [Fejes Tóth (1964); Schattschneider (1978)]

K3. Symmetries of regular and semiregular polyhedra. Find all the symmetries of these shapes. Show that you have all the symmetries for each shape.

K4. Finite groups of symmetry in space. The cube has finitely many symmetries. What are the possible finite groups of symmetries and what are shapes that go with each? Experience with ideas and techniques from modern algebra would help here. [Yale (1968); Senechal (1990)]

L. Efficient Shapes

L1. Investigate shapes of houses, buildings, and containers for efficiency: which uses the minimum amount of material in its construction, which costs the least to make, which incorporates a minimum heat loss, which costs the least to use, etc. This is a very sketchy project idea; it needs more focus, development, more references. [Niven (1981); Bezouszka (1984)]

L2. Which regular and other polyhedra are "efficient"? On pp. 139–141 of Pólya there are a bunch of optimization problems for three-dimensional shapes, mostly polyhedra. For example: Of all double pyramids of fixed volume, which has the smallest surface area? Some of these problems can be solved using the theorem of the arithmetic and geometric means. Others might be solved by analogy to solutions of problems in the plane using the three-dimensional Isoperimetric theorem. For example, some shapes are "best" when they are inscribable in a sphere. Try these problems. Look for connections and patterns. [Pólya (1954), pp. 139–141, 189; Fejes Tóth (1964), Goldberg (1935)]

L3. Efficient spherical polygons. Of all spherical polygons of the same area and same number of sides, which has least perimeter? [Fejes Tóth (1964), p. 213ff]

L4. Distributing *n* points on a sphere, I. How should you distribute *n* points on the sphere so that the least distance between the points should be as great as possible? What is the analogous problem on the circle and its solution? [Fejes Tóth (1964), p. 227ff]

L5. Distributing *n* points on a sphere, II. A problem that is possibly connected to L4: If you have *n* points arranged on a circle, then the convex hull has maximal area when the points are equally spaced as the vertices of a regular *n*-gon. What about the three-dimensional problem: Of all arrangements of *n* points on a sphere, which ones have convex hulls of maximum volume? If there is a regular solid with *n* vertices, then shouldn't that be the answer for that case? (The **convex hull** of a set *S* is the smallest convex set containing *S*.) [Pólya (1954), pp. 188–189]

M. Packing

M1. Diamond cutting. How a diamond is cut may be related to the geometry of the diamond molecule and how the molecules arrange themselves in aggregate. What are these geometric configurations? Sphere packing may be related to this. What is the geometry of other crystalline forms of carbon? What does a jeweler do with a diamond? What does it have to do with the geometry of the diamond's structure? [Pauling and Hayward (1964); Lenzen (1983); Pierson (1993)]

M2. Random patterns in two dimensions. More on random packing of circles and cells and other phenomena, such as metal grain boundaries. What do the patterns have in common? How can they be classified or analyzed? [Weaire (1984)]

M3. Random packing of spheres, I. Fejes Tóth (1964, p. 306) describes experiments where you drop grains of shot into a vessel with a curved bottom. This gives you a loose random packing. Shake the vessel and you get a closer packing. Shake the vessel for a while and you get a close random packing. Try these out and repeat several times. What densities do you get? How do the densities compare with cubic close packing? The fact that the densities of the loose and close random packings differ essentially was known in antiquity. Luke 6:38 says "Give, and it shall be given unto you; good measure, pressed down, and shaken together, and running over shall men give into your bosom. For with the same measure that ye mete withal it shall be measured to you again." Can you support the numbers you get with arguments? [Coxeter p. 409; Stevens]

M4. Random packing of spheres, II. If you randomly pack spheres then compress them, what do the compressed shapes look like? (For cubic close packing, they're rhombic dodecahedra.) Presumably the shapes are polyhedra. What kinds? How many faces do they have? What kinds of faces do the polyhedra have? You might want to figure out a three-dimensional Euler's formula for dividing a large polyhedron into a smaller one, then try to mimic the argument for closest circle packing in text that uses Euler's formula in the plane. Here's a quote from Peter Stevens (1974, p. 188) that you might want to reconcile with your results and compare with the quote from Hale's *Vegetable Staticks* in the text: "no packing of [irregular] polyhedra with pentagonal faces can uniformly fill space." Do the shapes from a loose random packing differ from those in a close random packing? For the experiments, you might want to try plasticene spheres rolled in powdered chalk. [Coxeter (1969), p. 409]

M5. More on the design of honeycombs. Suppose that a hexagonal tube is capped at one end with three rhombi and that the other end is open. Find the angle between the tube center line and end faces that will minimize the surface area of the tube-with-cap for a given volume. Compare result with what we know about bee cells. Fejes Tóth has found a different kind of shape with which to cap a bee cell which meets all the requirements of such a cap but uses less wax. What is this shape? Make a model of it. [Fejes Tóth (1964b); Peressini (1980)]

M6. Sphere packing and coding theory. Make sense out of the following quote from Sloane: "The study of sphere packings in *n* dimensions has been recognized for some time as being mathematically equivalent to the design of a finite set of digitally encoded messages that do not waste power or cause confusion in transmission." [Sloane (1984)]

✳ *Resources*

Note: The issues of the journal, *Structural Topology*, contain many articles related to the ideas above, and more. The Internet's World Wide Web is also a place to look for ideas. You might want to start by looking at this book's Web page at http://www.wiley.com/college/gay.

Almgren and Taylor, "The geometry of soap films and soap bubbles," *Scientific American*, July 1976.

Anno, Mitsumasa, and Masaichiro Anno, *Anno's magical ABC: An anamorphic alphabet*. New York: Philomel Books, 1980.

Baglivo, J. A., and Graver, J. E., *Incidence and symmetry in design and architecture*. New York: Cambridge University Press, 1983.

Ball, W. W. Rouse, and Coxeter, H. S. M., *Mathematical recreations and essays*. New York: Macmillan, 1962.

Banchoff, Thomas F., *Beyond the third dimension: Geometry, computer graphics and higher dimensions*. New York: Scientific American Library, Distributed by W. H. Freeman, 1990.

Banchoff, Thomas F., *The hypercube: projections and slicing*. Film. Distributed by International Film Bureau, Chicago, 1978.

Bassein, Richard S., "The length of the day," *American Mathematical Monthly*, December 1992, 917–921.

Beard, Robert Stanley, *Patterns in space*. Palo Alto, CA: Creative Publications, 1973.

Beck, A., Bleicher, M. N., and Crowe, D. W., *Excursions into mathematics*. New York: Worth Publishers, 1970.

Bern, Marshall W., and Ronald L. Graham, "The shortest network problem," *Scientific American*, January 1989, 84–89.

Bezuszka, S., *Word problems for maxima and minima*. Boston College Press. 1984.

Boltianskii, V. G., *Equivalent and equidecomposable figures*. Boston: Heath, 1963.

Broman, Arne, and Lars Broman, "Museum exhibits for the conics," *Mathematics Magazine*, June 1994, 206–209.

Chakravarty, N., et al., "One story building as tensegrity frameworks." *Structural Topology,* no. 12 (1986).

Connelly, Robert, "The rigidity of polyhedral surfaces," *Mathematics Magazine,* November 1979, 275–283.

Courant, R., and Robbins, H., *What is Mathematics? An elementary approach to ideas.* New York: Oxford University Press, 1973.

Coxeter, H. S. M., *Introduction to geometry.* New York: Wiley, 1969.

Cromwell, Peter R., "Kepler's work on polyhedra," *Mathematical Intelligencer*, vol. 17, no. 3 (1995), 23–33.

Cundy, H. M., and Rollett, A. P., *Mathematical models.* Oxford: Clarendon Press, 1961.

Drucker, Daniel, "Reflection properties of curves and surfaces," *Mathematics Magazine,* June 1992, 147–157.

Edmondson, Amy C., *A Fuller explanation: The synergetic geometry of R. Buckminster Fuller.* Boston: Birkhauser, 1987.

Fejes Tóth, L., *Regular figures.* New York: Macmillan, 1964.

Fejes Tóth, L., "What the bees know and what they do not know," *Bulletin of the American Mathematical Society,* vol. 70 (1964), 468–481.

Fusè, Tomoko, *Unit origami: multidimensional transformations.* New York: Japan Publications, 1990.

Gay, D., *Three-dimensional kaleidoscopes.* Unpublished report. Department of Mathematics, University of Arizona.

Goldberg, M. "The isoperimetric problem for polyhedra," *Tohoku math jour.*, 40, pp. 226–36, 1935.

Grünbaum, Branko, and G. C. Shephard, *Tilings and patterns.* New York: W. H. Freeman, 1987.

Grünbaum, Branko, and G. C. Shephard, "Rigid plate frameworks," *Structural Topology,* no. 14 (1988), 1–8.

Heath, Thomas Little, *The thirteen books of Euclid's Elements.* New York: Dover, 1956.

Heath, Thomas Little, *The works of Archimedes.* New York: Dover, 1953.

Hecht, Eugene, *Optics.* Reading, MA: Addison-Wesley, 1987.

Hilbert, David, and S. Cohn-Vossen, *Geometry and the imagination.* New York: Chelsea Publishing Co., 1952.

Hildebrandt, S., and A. Tromba, *Mathematics and optimal form.* New York: Scientific American Library. Distributed by W. H. Freeman, 1985.

Holden, Alan, *Shapes, space and symmetry.* New York: Columbia University Press, 1971.

Hopley, Ron, "Nested Platonic solids: a class project in solid geometry." *Mathematics Teacher,* May 1994, 312–318.

Isenberg, Cyril, *The science of soap films and soap bubbles.* Clevedon: Tieto Ltd., 1978.

Jennings, George A., *Modern geometry with applications.* New York: Springer-Verlag, 1994.

Kaplansky, Irving, *Hilbert's Problems.* Notes. Department of Mathematics, University of Chicago, 1977.

Kappraff, Jay, *Connections: The geometric bridge between art and science.* New York: McGraw-Hill, 1991.

Kasahara, Kunihiko, and Toshie Takahama, *Origami for the connoisseur.* New York: Jan Publications, Inc., 1987.

Katz, Victor, *A history of mathematics: an introduction.* New York: Harper Collins, 1993.

Kitrick, Christopher J., "Geodesic domes," *Structural Topology*, no. 11 (1985), 15–20.

Layzer, David, *Constructing the universe.* New York: Scientific American Library, Distributed by W. H. Freeman, 1984.

Lenzen, Godehard, *Diamonds and diamond grading.* Boston: Butterworth & Co. Ltd, 1983.

Lines, L., *Solid geometry*. New York: Dover, 1965.

Loeb, Arthur L., *Space structures- their harmony and counterpoint*. Reading, MA.: Addison-Wesley, 1976.

McCrory, Clint, *Scissors congruence and Dehn's theorem*. Video. Department of Mathematics, California State University at Humbolot, 1986.

McMahon, Thomas A., and John Tyler Bonner, *On size and life*. New York: Scientific American Library. Distributed by W. H. Freeman, 1983.

Maesumi, Mohsen, "Parabolic mirrors, elliptic and hyperbolic lenses," *American Mathematical Monthly*, June–July 1992, 558–560.

Moore, C., "Contemporary conic sections: the ellipse and the lithotripter," *COMAP consortium newsletter*, November 1986.

Morgan, Frank, "Mathematicians, including undergraduates, look at soap bubbles," *American Mathematical Monthly*, April 1994, 343–351.

Nelson, David R., and B. I. Halperin, "Pentagonal and icosahedral order in rapidly cooled metals," *Science*, vol. 229 (July 1985), 233–238.

Newman, James R., *The world of mathematics*. New York: Simon and Schuster, 1956.

Niven, Ivan, *Maxima and minima without calculus*. Washington D.C.: Mathematical Association of America, 1981.

O'Daffer, Phares, and Stanley R. Clemens, *Geometry: an investigative approach*. Menlo Park, CA: Addison-Wesley, 1997.

Pauling, Linus, and Roger Hayward, *The architecture of molecules*. San Francisco: Freeman, 1964.

Pearce, Peter and Susan, *Polyhedra primer*. New York: Van Nostrand Reinhold, 1978.

Peressini, Anthony L., "The design of honeycombs," *Modules and Monographs in Undergraduate mathematics and its Applications Projects (UMAP)*, Unit 502, EDC/UMAP, 55 Chapel St., Newton, MA 02160, 1980.

Pierson, Hugh O., *Handbook of carbon, graphite, diamond and fullerenes*. Park Ridge, NJ: Noyes Publications, 1993.

Pólya, Georg, *Induction and analogy in mathematics, Vol I of Mathematics and plausible reasoning*. Princeton, NJ: Princeton University Press, 1954.

Ranucci, E. R., and J. L. Teeters, *Creating Escher-type drawings*. Palo Alto: Creative Publications, 1977.

Rickey, V. Frederick, "How Columbus encountered America," *Mathematics Magazine*, October 1992, 219–225.

Row, T. Sundara, *Geometric exercises in paper folding*. New York: Dover, 1966.

Saunders, Martin, "Buckminsterfullerine: The inside story," *Science*, vol. 253 (July 1991), 330–331.

Schattschneider, D., "The plane symmetry groups: Their recognition and notation," *American mathematical monthly*. June 1978, 439–450.

Schattschneider, D., and Wallace Walker, *M. C. Escher Kaleidocycles*. Rohnert Park, CA: Pomegranate Art Books Inc., 1977.

Schattschneider, Doris, *Visions of symmetry: Notebooks, periodic drawings, and related work of M. C. Escher*. New York: W. H. Freeman, 1990.

Schmalz, T. G., et al., "Elemental carbon cages," *Journal of the American Chemical Society*, vol. 110 (1988), 1113–1127.

Senechal, Marjorie, "Which tetrahedra fill space?" *Mathematics Magazine,* November 1981, 227–243.

Senechal, Marjorie, and George M. Fleck, *Shaping space: a polyhedral approach*. Boston: Birkhauser, 1988a.

Senechal, Marjorie, and George M. Fleck, *A workbook of common geometry*. Notes. Northampton, MA: Smith College, 1988b.

Senechal, Marjorie, "Finding the finite groups of symmetry of the sphere," *American Mathematical Monthly*, April 1990, 329–335.

Senechal, Marjorie, et al., *Penrose tiles! Notes*. Northampton, MA: Smith College, 1996.

Seymour, Dale, *Introduction to line designs*. Palo Alto: Dale Seymour Publications, 1992.

Shubnikov, A. V., and V. A. Koptsik, *Symmetry in science and art*. New York: Plenum Press, 1974.

Sloane, N. J. A., "The packing of spheres," *Scientific American*, January 1984, 116–125.

Steen, Lynn, ed., *For all practical purposes: introduction to contemporary mathematics*. New York: W. H. Freeman, 1988.

Steinhaus, Hugo, *Mathematical snapshots*. New York: Oxford University Press, 1950.

Stevens, Peter S., *Patterns in nature*. Boston: Little, Brown, 1974.

Thomas, David Emil, "Mirror images," *Scientific American*, December 1980, 206–228.

Thompson, D'Arcy Wentworth, *On growth and form*. Cambridge: University Press, 1961.

Tokarsky, George W., "Polygonal rooms not illuminable from every point," *American Mathematical Monthly*, December 1995, 867–879.

Tong, Jungcheng, and Yap S. Chua, "The generalized Fermat's point," *Mathematics Magazine*, June 1995, 214–215.

van Heldon, Albert, *Measuring the universe: Cosmic dimensions from Aristarchus to Halley*. Chicago: University of Chicago Press, 1985.

Wagon, Stan, "Why December 21 is the longest day of the year," *Mathematics Magazine*, December 1990, 307–311.

Walker, Jearl, "The amateur scientist: Mirrors make a maze so bewildering that the explorer must rely on a map," *Scientific American*, June 1986, 120–126.

Waugh, Albert E., *Sundials: Their theory and construction*. New York: Dover, 1973.

Weaire, D., and N. Rivier, "Soap, cells and statistics—random patterns in two dimensions." *Contemporary Physics*, vol. 25, no. 1 (1984), 59–99.

Whitt, Lee, "The standup conic presents: The parabola and applications," *The UMAP Journal*, vol. III, no. 3 (1982), 285–313.

Whitt, Lee, "The standup conic presents: The hyperbola and applications," *The UMAP Journal*, vol. V, no. 1 (1984), 9–21.

Whitt, Lee, "The standup conic presents: the ellipse and applications," *The UMAP Journal*, vol. IV, no. 4 (1983), 157–183.

Yale, Paul B., *Geometry and symmetry*. San Francisco: Holden-Day, 1968.

✳ *Communicating Project Results*

Think of your project as the response to a question. For example, "What is the best shape for such-and-such?" As in solving any problem, the response will occur in stages, such as

• An exploration

• Gathering data

• Search for a pattern

• Working out examples

• Making a model

• Coming up with an answer to the question, or a partial answer, or an answer to a similar or related question

• Making a convincing argument explaining the answer

When your project involves an expository component, you will likely find material written about your problem. Read and digest it, knowing that you will be explaining it to someone, very effectively. Put together the material in a new way. Relate it to things we've done in the course (or to previous experience with geometry). Make a model (or two) that illustrates the material.

Even Though the Material May Not Be New to the World at Large, It Will Be New to You. Wrestling with the ideas, making examples, building models and putting together an exposition of them always produces a new prospective on the original material and frequently produces new proofs and insights. I have chosen the topics above to maximize the possibilities of making connections with material in course, possibilities for models, possibilities to put materials together in new ways appropriate for you.

For example, in a project on the Buckyfullerine molecule, students in the team struggled with recent journal articles. They looked at models. They drew on their experience with polyhedra to make sense of the mathematical aspects of molecule possibilities.

After you have found an answer (or answers) you are happy with, then you must work on communicating the results of these earlier stages. This communicative aspect of the project could have one or more of the following parts:

- **A display** This will take the form of a display (Science Fair style) which in turn will include a summary of your results: the question to be answered, your plan of attack, ideas you used/developed, and a summary of your results. This aspect of the project should convey in a visual way an overall feeling for the project (its results and its procedures), without providing too many details. This is a good place to use models, illustrations, diagrams, and charts. There is more on the display in the next section.

- **An oral presentation** This is an occasion for you (and your team) to introduce to others the results of your project and to point out important features of the display (if there is one). The oral presentation should communicate effectively a few of the main ideas of the project's topic, its flavor, its interest to other mathematicians, its applications in the real world, how it relates to the topics encountered in the rest of the book, and perhaps an esoteric detail or two.

Here is where you can use a cute trick or gadget. You want to convey briefly the flavor of the project—an aspect of its solution that's easy to communicate, a connection with other problems the listener will know, the significance of the problem. You want to strike a balance between the trivial and the complex: On the one hand, you don't want to spend time belaboring the obvious; on the other, you do not want to subject your listener to an involved proof. (In both cases the listener would be bored.) If listeners want to know more than you are able to include in your presentation, let them ask questions. Your presentation will be more successful if you let them participate: pick up on where they are puzzled, on what their interests are. You must remember that you have been living with your topic and are an expert on it. Your listener is not an expert. Don't overwhelm her with jargon. What you think is trivial (relative to the project) might be obscure

to the uninitiated. Practice your presentation. Time it. Practice it again. Make sure you have all your props ready and in order. (If you have to fumble around during your presentation for a piece of paper or a model, you will loose precious time and also your audience.)

- **A paper** Like any other paper it should be an exposition of the results of the project; it should describe the ideas, procedures, and theorems the team used to answer the question/solve the problem; it should include the answer(s) and written arguments justifying them. If your project involves a display and/or oral presentation, the paper is the opportunity for you (and your team) to expand on the broad ideas and impressions indicated in the display/oral presentation and provide more of the details to back up what was claimed there. In the paper, you put the pieces together and show how they are connected. The paper can utilize as many pieces of the display or presentation as you wish, but there should be no loose ends.

Make sure that your reader knows what question (or set of questions) you are answering. Describe how you (or others) explored the terrain in looking for answers. Provide some of the examples you worked out. Or create new examples useful for demonstrating your lines of attack or for pointing out possible blind alleys. Again, provide pictures, diagrams, and tables to help illustrate your ideas and summarize your results.

You want the arguments to be valid. You want what you write to make sense. If you quote source materials, make sure you understand what those sources say. Document your sources.

✳ *Nitty-Gritty on Putting Together an Effective Display*

Get a Backdrop for Your Display.   The easiest thing to do here is to buy a coated, corrugated cardboard triptych, about 36 in high by 48 in wide, which many office supply stores sell. Alternatively, you can buy a piece of foam board for around $5. Then, with knife and a metal straight edge, score it twice to make a triptych which will stand up on its own. (See pictures below.)

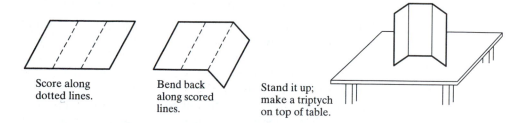

Score along dotted lines.   Bend back along scored lines.   Stand it up; make a triptych on top of table.

Choose a Title for the Project.   The title should be a question which your project answers. It should be immediately understandable to any person viewing your display. Place the title in a prominent place on the triptych.

Outline Main Aspects of the Project. Here is a sample outline:

- Detailed statement of problem
- Significance of problem
- Procedures used (steps taken) to solve problem
- Results obtained
- Arguments used
- Illustrative examples
- Descriptions of models accompanying the display
- Conclusions

You could think of this as an outline for the project's paper as well. Projects will differ. Depending on the project, you may want to delete or add to the list above.

Decide on Windows. Plan on having six to ten "windows of information" to attach to the display board. Each "window" should correspond to an item in your outline and would be labeled by the outline heading. Each window should be as much of a visual display of information as possible, employing diagrams, photos, and graphs, and using only brief verbal descriptions.

Make It Spiffy. To make your display attractive, you may want to use large, computer generated letters or stick-on letters for the title. To accompany the title, you may want to have a single, large, catchy, and representative picture or diagram. You may want to place each "window" on a background of colored paper to make it stand out. You want your display to draw in viewers, and get them interested and intrigued. You are selling your project. You want the viewer to become curious about it and to ask questions.

✳ *Project Evaluation*

Here are criteria (the four C's) for evaluating every aspect of your project—oral presentation, display, or paper.

Clarity. Is the project topic (or problem) clear? Are project goals—how the topic is to be dealt with, what kind of problem resolution is sought—clear? Is the presentation carefully organized? Can an outsider see how the various pieces of the project are connected?

Completeness. Have all reasonable lines of inquiry been investigated? If some lines have been used and not others, have the choices been explained? (For example, "There was not enough time to pursue such and such a line of reasoning," "Such and such approach appeared to be too difficult,". . . .) Have connections been made between the project and other areas of study, especially areas covered in the course? Is the project problem analogous to a problem discussed in the course? Is it a generalization? Does the project shed light on a problem seen before? Is the solution to the project problem similar to the solution of a problem seen before? What are related problems which need further study? Why is the project problem important? Does its solution have applications to some part of mathematics? Is its solution useful to another discipline? To the real world? Did the problem remain unsolved for a long time? Does the project incorporate effective uses of models, charts or diagrams where appropriate? Does the project incorporate effective use of technology where appropriate?

Correctness.   Are all parts of the project mathematically correct? Are definitions correctly stated? Are the arguments correct? Have all aspects of the project been communicated effectively in correct English? Are outside sources acknowledged? Is there an appropriate bibliography?

Creativity.   Does the team use an original approach to solving a problem? Are there are unusual ways of putting ideas together? Are the means of communicating the project's topic or its solution new? Does the project include the effective use of a clever model? Has the project used technology (computers, videos, overheads, etc.) creatively?

✳ *The Bottom Line*

In the end, what should you to get out of the project?

- Research experience: Choosing a geometric topic, working independently to understand a body of material related to it, and making the related ideas hang together.
- The experience of having a real context (more or less self-created) out of which to write mathematical ideas, put together an effective visual display, and make an oral presentation.
- The experience of communicating your insight to others.
- Making models and appreciating their value, for explanation, for illustration and for understanding.
- Experience working with a team—learning from and communicating with your peers.
- Sense of pride in a job well done.
- Passion. Really getting into something you're interested in.

Photo Credits

page 136—from *The Architecture of Molecules* by Linus Pauling and Roger Haywood. Copyright © 1964. Used by permission of W. H. Freeman and Company.

page 138—from *Shaping Space: A Polyhedral Approach*, Marjorie Senechal and George Fleck, eds. (Boston: Birkhäuser, 1988). Used by permission.

page 177—Gary Larson, *The Far Side*. Copyright © Universal Press Syndicate. Used by permission.

page 279—(center) Reproduced by permission from *Art of a Vanished Race: The Mimbres Classic Black-on-White* (Dillon Tyler Publishers, 1975)

page 342—from *Patterns in Nature* by Peter Stevens. Copyright © 1974 by Peter S. Stevens. Used by permission of Little, Brown and Company.

page 343—(left and center) from *Patterns in Nature* by Peter Stevens. Copyright © 1974 by Peter S. Stevens. Used by permission of Little, Brown and Company.

page 343—(right) from *On Growth and Form* by D'Arcy Wentworth Thompson. (New York: Dover Publications, Inc., 1992) Used by permission.

page 378—from *Shaping Space: A Polyhedral Approach,* Marjorie Senechal and George Fleck, eds. (Boston: Birkhäuser, 1988). Used by permission.

page 380—from *Shaping Space: A Polyhedral Approach*, Marjorie Senechal and George Fleck, eds. (Boston: Birkhäuser, 1988). Used by permission.

Index